Geir Legreid

OVOCs in the Swiss Boundary Layer: Measurement and Source Allocation

Geir Legreid

OVOCs in the Swiss Boundary Layer: Measurement and Source Allocation

Seasonal measurement campaigns at urban and remote sites with a double adsorbent sampling system coupled to GC-MS

Südwestdeutscher Verlag für Hochschulschriften

Impressum / Imprint
Bibliografische Information der Deutschen Nationalbibliothek: Die Deutsche Nationalbibliothek verzeichnet diese Publikation in der Deutschen Nationalbibliografie; detaillierte bibliografische Daten sind im Internet über http://dnb.d-nb.de abrufbar.
Alle in diesem Buch genannten Marken und Produktnamen unterliegen warenzeichen-, marken- oder patentrechtlichem Schutz bzw. sind Warenzeichen oder eingetragene Warenzeichen der jeweiligen Inhaber. Die Wiedergabe von Marken, Produktnamen, Gebrauchsnamen, Handelsnamen, Warenbezeichnungen u.s.w. in diesem Werk berechtigt auch ohne besondere Kennzeichnung nicht zu der Annahme, dass solche Namen im Sinne der Warenzeichen- und Markenschutzgesetzgebung als frei zu betrachten wären und daher von jedermann benutzt werden dürften.

Bibliographic information published by the Deutsche Nationalbibliothek: The Deutsche Nationalbibliothek lists this publication in the Deutsche Nationalbibliografie; detailed bibliographic data are available in the Internet at http://dnb.d-nb.de.
Any brand names and product names mentioned in this book are subject to trademark, brand or patent protection and are trademarks or registered trademarks of their respective holders. The use of brand names, product names, common names, trade names, product descriptions etc. even without a particular marking in this works is in no way to be construed to mean that such names may be regarded as unrestricted in respect of trademark and brand protection legislation and could thus be used by anyone.

Coverbild / Cover image: www.ingimage.com

Verlag / Publisher:
Südwestdeutscher Verlag für Hochschulschriften
ist ein Imprint der / is a trademark of
AV Akademikerverlag GmbH & Co. KG
Heinrich-Böcking-Str. 6-8, 66121 Saarbrücken, Deutschland / Germany
Email: info@svh-verlag.de

Herstellung: siehe letzte Seite /
Printed at: see last page
ISBN: 978-3-8381-3363-8

Zugl. / Approved by: Zürich, ETH, Diss., 2006

Copyright © 2012 AV Akademikerverlag GmbH & Co. KG
Alle Rechte vorbehalten. / All rights reserved. Saarbrücken 2012

Content

Abstract..3
1 Introduction..7
 References..9
2 The role of organic gases in tropospheric chemistry.......................................12
 2.1 Tropospheric photochemistry...12
  2.1.1 Sources of hydroxyl radicals...12
  2.1.2 Mechanisms of VOC oxidation...13
   2.1.2.1 Alkanes...13
   2.1.2.2 Alkenes...14
   2.1.2.3 Alkynes...16
   2.1.2.4 Aromatics..16
  2.1.3 Destruction mechanisms for the OVOCs....................................16
 2.2 Daytime photo oxidant formation..17
  2.2.1 Photostationary state..17
  2.2.2 Ozone formation in polluted environments.................................18
  2.2.3 Photochemical Ozone Creation Potential (POCP) and tropospheric lifetimes..19
 2.3 Sources of OVOCs...20
 2.4 Night-time oxidation..22
 References..23
3 Measurement of organic gases in ambient air and instrument development............26
 3.1 Introduction..26
 3.2 GC systems for OVOC analysis...26
 3.3 Instrument development..27
  3.3.1 Tubing...28
  3.3.2 Water removal..28
   3.3.2.1 Cold trap..29
   3.3.2.2 Adsorbtive removal of water.......................................33
   3.3.2.3 Chromosorb W AW with lithium chloride.......................34
   3.3.2.4 Summary..36
  3.3.3 Sampling artefacts by ozone..36

			3.3.3.1 Potassium iodide ozone scrubber..................................38
			3.3.3.2 Ozone removal by NO titration.....................................42
		3.3.4	Adsorbent material...43
		3.3.5	Separation..45
		3.3.6	Detection by GC-MS and calibration.....................................45
	3.4	Summary of the developed OVOC instrument....................................49
	3.5	Intercomparison at the SAPHIR smog chamber.................................. 50
	3.6	Measurement campaigns..50
	3.7	Blank issues...51
	References..52
4	OVOC measurements in the Gubrist highway tunnel... 54
5	OVOC measurements in Zürich...73
6	OVOC measurements at Jungfraujoch... 99
7	Conclusions and outlook..120

Abbreviations...124

Appendix I	Measurements in the Swiss alpine village Roveredo............................ 125
Appendix II	Gas standard mixtures used in the experiments....................................127

Acknowledgements.. 128

Abstract

Oxygenated Volatile Organic Compounds (OVOCs) play an important role in tropospheric chemistry. Such compounds are, for example, important intermediates in the oxidation of many primary pollutants and precursors for peroxyacylnitrates (PANs). PANs are carriers of reactive nitrogen potentially releasing free radicals at remote sites. In other words, the OVOCs influence the oxidizing capacity and the ozone-forming potential of the atmosphere. They may also contribute significantly to the formation of secondary organic aerosols (SOAs). Knowledge of their distribution and sources is still restricted to mostly short-time measurements of few compounds.

In this study 21 OVOCs and selected non-methane hydrocarbons (NMHCs) were measured with a recently developed double adsorbent sampler coupled to a gas chromatograph-mass spectrometer (GC-MS). Measured compounds were aldehydes and ketones, key intermediates of tropospheric chemistry as well as primary anthropogenic and biogenic compounds; alcohols, emitted by both anthropogenic and biogenic sources; and ethers and esters, which are mostly emitted by anthropogenic sources. Furthermore, selected anthropogenic and biogenic NMHCs were measured. Measurement locations were a highway tunnel (near Zürich), an urban background station in Zürich, a remote Alpine site and a village in the Swiss Alps (Roveredo), in which the air was highly influenced by wood burning emissions in winter.

The OVOC measurements in the highway tunnel were used to estimate the contribution of the Swiss vehicle fleet to OVOC emissions. Ethanol was the most abundant compound found in this study with an emission factor (EF) of 10 mg/km. This compound was not only related to exhaust emissions, but also to the use of window wiper fluids. In total, the OVOCs represented 54 % of the measured volatile organic compounds (VOCs) from mobile sources. The measurements indicated that OVOCs were mainly emitted by the heavy-duty vehicles (HDV), whereas the light-duty vehicles (LDV) dominated the emissions of the NMHCs. The comparison with earlier campaigns at the same site confirmed the large decrease of organic exhaust emissions under highway conditions, due to steady improvements of vehicle technology.

The measurement campaigns in Zürich were performed in order to gather information about the sources of the OVOCs in Switzerland. These were the first data collected for many of the OVOCs in the Swiss boundary layer, and are therefore unique in this respect. Also in Zürich, ethanol was the dominating compound measured throughout all seasons. Its anthropogenic origin was indicated by higher mixing ratios in winter than in summer, which was also the case for known anthropogenic pollutants like benzene and acrolein. On the other hand, compounds with additional biogenic sources like methanol, acetone and isoprene had higher levels during summer. Local

Abstract

sources were estimated to contribute to 40 % and 49 % to the OVOC mixing ratios in summer and winter, respectively. Combustion was responsible for about 75 % of these local sources independent of the season. About 50% of both the OVOC and NMHC levels in Zürich were explained by the regional background, which included regional biogenic and anthropogenic sources in addition to secondary production. From the calculation of the incremental ozone production, it was estimated that the OVOCs explained 40 % of the total VOC ozone production. Local OVOC sources were responsible for 16 %.

The campaigns at the high-alpine station Jungfraujoch aimed at a climatological description of OVOCs within the free troposphere above Europe and at a European source allocation. At this site acetone, methanol and acetaldehyde were the most abundant OVOCs, being responsible for 82 % of the measured VOCs in summer and 51 % in fall. The measured mixing ratios of these compounds were generally in accordance with other studies from remote locations.

Source regions for the pollutants at Jungfraujoch were estimated from measurement days with influence from the polluted boundary layer (PBL) by applying a statistical trajectory model. The mainly anthropogenic compounds ethanol, ethyl acetate, butane and benzene had two main source regions; southern Germany and northern Italy, both heavily populated and industrialized areas. For the two industrial solvents methyl acetate and butanone the main source region was solely northern Italy. Methanol and acetone, compounds which also have large biogenic sources, had their main contribution from northern Italy as well. This is probably due to the higher biogenic activity south of the Alps compared to the north.

Zusammenfassung

Oxygenierte flüchtige Kohlenwasserstoffe (OVOCs) spielen eine wichtige Rolle in der Troposphärenchemie. Diese Verbindungen treten als Zwischenprodukte in der Oxidation von vielen primären Luftschadstoffen auf, und sind Vorläufersubstanzen für Peroxyacylnitrate (PANs). PANs sind Trägersubstanzen für reaktive Stickoxide, die in entfernten Gebieten freie Radikale freisetzen können. Die OVOCs beeinflussen daher die Oxidationskapazität und Ozonproduktion in der Atmosphäre. Sie können auch eine wichtige Rolle bei der Bildung sekundärer organischer Aerosole (SOA) spielen. Das heutige Wissen über die Verbreitung und Quellen der OVOCs ist beschränkt, da bisherige Messungen meistens nur für kurze Zeiträume und für wenige Substanzen durchgeführt wurden.

In dieser Studie wurden mit einem neuartigen System Luftproben für die Analyse von 21 OVOCs und ausgewählte Nichtmethan-Kohlenwasserstoffe (NMHCs) gesammelt. Die Proben wurden mit einem Gas Chromatograph-Massenspektometer (GC-MS) analysiert. Die analysierten Verbindungen umfassten 1. Aldehyde und Ketone, sowohl wichtige Zwischenprodukte der Troposphärenchemie als auch primäre anthropogene und biogene Verbindungen; 2. Alkohole, emittiert von anthropogenen und biogenen Quellen; und 3. Ether- und Esterverbindungen, die meist anthropogener Herkunft sind. Zusätzlich wurden ausgewählte anthropogene und biogene NMHCs gemessen. Die Messungen wurden in einem Autobahntunnel, an einer städtischen Hintergrundstation (Zürich), auf einer hochalpinen Station (Jungfraujoch) und an einer Mess-Stelle einer Siedlung im Alpenraum (Roveredo), die durch große Emissionen von Holzverbrennung im Winter gekennzeichnet ist, durchgeführt.

Durch die Messungen der OVOCs im Autobahntunnel wurden die Beiträge des Straßenverkehrs an deren Emissionen quantifiziert. Ethanol war die Verbindung mit dem höchsten Mischungsverhältnis und einem Emissionsfaktor von 10 mg/km für die Gesamtflotte. Diese Verbindung wird nicht allein durch die Verbrennungsmotoren emittiert, sondern ist auch ein Hauptbestandteil der Scheibenwischerflüssigkeit. Insgesamt haben die OVOCs 54 % der gemessenen Kohlenwasserstoffe des Straßenverkehrs erklärt. Die Messungen haben gezeigt, dass die OVOCs hauptsächlich von LKWs (Lastkraftwagen) emittiert werden, hingegen die NMHCs vor allem von PKWs (Personenkraftwagen). Im Vergleich zu früheren Messungen im gleichen Tunnel sind die Emissionen der VOCs in den letzten Jahren aufgrund der ständigen Verbesserung der Kraftfahrzeugtechnologie kontinuierlich zurückgegangen.

Messungen an der städtischen Hintergrundstation in Zürich wurden durchgeführt um Informationen über die OVOC Quellen in der Schweiz zu erhalten. Für einige der gemessenen

Zusammenfassung

OVOCs sind dies die ersten Daten in der Schweizer Grenzschicht und stellen somit einen einzigartigen Datensatz dar. Wie im Tunnel, war Ethanol auch in der Umgebungsluft von Zürich die Verbindung mit der höchsten Konzentration. Die höhere Konzentration von Ethanol im Winter im Vergleich zu Sommer weist auf eine anthropogene Herkunft hin. Dies gilt auch für andere anthropogene Luftschadstoffe wie Benzol und Acrolein. Die Verbindungen mit zusätzlichen biogenen Quellen wie Methanol, Acetone und Isopren zeigten höhere Werte im Sommer. Lokale Quellen hatten einen Beitrag von 40 % bzw. 49 % zu dem OVOC Konzentrationen im Sommer und im Winter. Ungefähr 50 % der OVOC und NMHC Konzentrationen in Zürich wurden dem regionalen Hintergrund zugeschrieben, der biogene, anthropogene und auch sekundäre regionale Quellen beinhaltete. Bei der Berechnung der Ozonproduktion wurde ein Beitrag der OVOCs von 40 % der gesamten Ozonproduktion der Kohlenwasserstoffe abgeschätzt. Dabei waren lokale OVOC Quellen für 16 % der Ozonbildung verantwortlich.

Das Ziel der Messungen auf dem Jungfraujoch war eine klimatologische Beschreibung der OVOCs in der freien Troposphäre und eine Abschätzung deren europäischen Quellen. Auf der hochalpinen Station waren Aceton, Methanol und Acetaldehyd die Substanzen mit den höchsten Konzentrationen. Diese Verbindungen hatten einen Anteil an den gesamten gemessenen Kohlenwasserstoffen von 82 % im Sommer und 51 % im Herbst. Die gemessenen Werte waren im Allgemeinen in guter Übereinstimmung mit anderen Studien von wenig anthropogen beeinflussten Standorten.

Quellregionen für die Schadstoffe auf dem Jungfraujoch wurden an Tagen, an denen Luft aus der Grenzschicht durch Advektion an die Station herantransportiert wurde, durch ein statistisches Rückwärtstrajektorienmodell berechnet. Die anthropogene Verbindungen Ethanol, Ethylacetat, Butan und Benzol hatten zwei Haupt-Quellregionen in Süddeutschland und Norditalien. Die beiden Lösungsmitteln Methylacetat und Butanon (MEK) hatten ihre Hauptquellen in Norditalien. Methanol und Aceton, die beide große biogene Quellen haben, hatten ihre hauptsächlichen Emissionen in Norditalien, was sich aufgrund der höheren biologischen Aktivitäten südlich der Alpen erklären lässt.

1 Introduction

Air quality has been a concern of mankind since centuries. The ever growing human population and increasing industrialization, especially in developing countries, result in new issues to be addressed. The modern history of anthropogenic air pollution started with the industrial revolution at the end of the 18th century, where the large increase in the use of fossil fuel resulted in major smog episodes in the larges cities. Thus, in 1952 the population of London suffered under this "winter smog", which was mainly caused by coal-burning stoves. About 4,000 people died within a few days from the elevated concentrations of soot and sulphur built up in a stagnant, foggy air mass trapped by a temperature inversion (Brimlecombe 1987). This problem has been overcome in the industrialized countries due to strong regulations on fuel characteristics (among others sulphur content), less polluting industry in populated areas, and better combustion and cleaning technology. Still, the industrialized countries have large problems with air pollution in urban areas due to ever increasing road traffic. In winter, high amounts of particulate matter from stationary combustion and diesel fuelled engines are made responsible for thousands of early deaths each year (Boldo et al. 2006), and in summer high observed ozone mixing ratios have adverse effects on crops, human health and materials (Heck et al. 1984; Tidblad et al. 2004; Bergin et al. 2005).

The photochemical smog was first discovered in the cities of the US West Coast, but was afterwards detected in nearly every urban environment during the warm season. Photochemical smog is due to the reaction of volatile organic compounds (VOCs) with nitrogen oxides ($NO_x = NO + NO_2$) which under solar radiation produces secondary pollutants like ozone, secondary organic aerosols (SOA), peroxyacylnitrates (PANs) and other oxidation products like organic acids, carbonyls etc. (Atkinson 2000). VOCs include all species of vapour phase organics in the atmosphere (Seinfeld 1998), which, on a global scale, are predominantly of natural origin (Fall 2003). Guenther et al (1995) estimated an annual global non-methane hydrocarbon (NMHC) flux of 1150 Tg yr^{-1} primarily composed of isoprene (44%) and monoterpenes (11%). Global anthropogenic NMHC emissions were rated to be 60-140 Tg yr^{-1}. Due to the detrimental effects of ozone on human health, animals and vegetation, most countries, as well as the World Health Organisation (WHO), have established goals for maximum allowable ozone concentrations in the air of about 100 ppb (averaged over a one-hour period). In Switzerland 60 ppb hourly mean value should not be exceeded more than once every year. Even if measures have been introduced, the given limits are regularly exceeded in numerous cities all over the world. (Jorquera et al. 1998; Guttikunda et al. 2005; Im et al. 2006). This is partly due to the complexity of the processes involving numerous chemical reactions, which make it difficult to introduce efficient reduction strategies.

1. Introduction

The production of tropospheric ozone depends on the availability of both NO_x and VOCs. Under high NO_x conditions (e.g. central urban environments), ozone production is limited by the availability and reactivity of the VOCs (VOC-limitation). Conversely, under low NO_x conditions (e.g. remote rural areas), ozone production is limited by the availability of NO_x (NO_x-limitation). The transition point between the VOC- and NO_x-limited ozone formation is determined by the sources and sinks of the radical intermediates involved (Sillman 1999) (see chapter 2.2.2). Ozone production in areas removed from primary pollution can be affected by long-range transport of precursors from more polluted areas, as well as by local natural sources of VOCs and NO_x.

SOA is formed by chemical reactions and gas-to-particle conversion of organic vapours emitted into the atmosphere. There are three possible pathways (Poschl 2005):

1) New particle formation. Semivolatile organic compounds (SVOCs) are formed from oxidation processes, and they take part in the nucleation and growth of new aerosol particles.

2) Gas-particle partitioning. The SVOCs are formed by gas-phase oxidation and are taken up by pre-existing aerosol or cloud particles

3) Heterogeneous or multiphase reactions. Formation of low-volatility or non-volatile organic compounds (LVOCs, NVOCs) by chemical reaction of VOCs or SVOCs at the surface or in the bulk of aerosol or cloud particles.

In the aerosol phase chemical reactions can produce numerous organic compounds, which due to the small size of the aerosol particles (< 0.1nm) can reach the most inner parts of our body (Nemmar et al. 2002).

To improve the air quality in those urban environments that are subject to photochemical smog, the amount of reactants, principally NO_x and fast-reacting VOCs (e.g. alkenes (see chapter 2)), emitted into the air must be reduced. Reduction measures of VOCs and NO_x have not resulted in the same reductions in ground ozone mixing ratio (Ordonez et al. 2005). This could be due to the influence from more polluted background air transported from other continents, which has been suggested by several research groups (Broennimann et al. 2000; Stohl 2001; Naja et al. 2003).

Switzerland has one of the densest networks for measurements of air pollutants in the world: the National Air Pollution Monitoring Network (NABEL). 16 stations measure NO_x, NO, NO_2, ozone and meteorological parameters continuously. NMHCs are measured at 4 stations:

Rigi (rural, pre-alpine region, 1031 m asl): C_2-C_9 NMHCs.

Zürich (urban background station, 410 m asl): C_2-C_9 NMHCs.

Dübendorf (urban background station, 433 m asl): aromatic VOCs.

1. Introduction

Jungfraujoch (remote station, 3578 m asl): C_2-C_7 NMHCs.

Until now oxygenated VOCs (OVOCs) were not regularly measured and represent a gap of knowledge. The reason is that the measurements of the OVOCs are not as straightforward as for the NMHCs. This is due the analytical problems involving detector response (FID), removal of sample humidity, low breakthrough volume for adsorbents and lack of stable low concentration standards. Only a low number of measurements of these compounds have been performed during campaigns of limited spatial and temporal coverage (Riemer et al. 1998; Apel et al. 2003; Singh et al. 2004; Schade and Goldstein 2006), and in Switzerland only short-term point measurements have collected data for these compounds (BUWAL 1999).

Since almost no OVOC measurements existed in Switzerland, a suitable instrument was constructed in this PhD thesis, and was thereafter used in several campaigns. The project was named "Emissions of Non-regulated Oxidised Volatile Organic Compounds into the Polluted Troposphere Analysed by Advanced GC-MS Technology" (ENOVO).

The thesis is separated into five chapters. In chapter 2 the tropospheric chemistry is briefly summarized as far as important for the PhD thesis. Chapter 3 describes the instrument development, which was the main focus during the first part of this work. In chapter 4 to 6 the measurement campaigns and their analysis are presented in the following manner:

- Road traffic measurements in the Gubrist tunnel close to Zürich for estimation of real-world emissions of the OVOCs from vehicle transport (Chapter 4).
- Seasonal campaigns in the urban environment of Zurich (Chapter 5).
- Seasonal campaigns at the high alpine site Jungfraujoch (Chapter 6).

Chapters 4 to 6 are presented in the form of three papers. Conclusions and suggestion for further work is presented in chapter 7. Furthermore, the measurements in a village in the Swiss Alps (Roveredo), in which the air was highly influenced by wood burning emissions in winter are described in appendix I.

References:

Apel, E. C., A. J. Hills, et al. (2003). "A fast-GC/MS system to measure C-2 to C-4 carbonyls and methanol aboard aircraft." Journal of Geophysical Research-Atmospheres **108**(D20): art. no.-8794.

Atkinson, R. (2000). "Atmospheric chemistry of VOCs and NOx." Atmospheric Environment **34**(12-14): 2063-2101.

Bergin, M. S., J. J. West, et al. (2005). "Regional atmospheric pollution and transboundary air quality management." Annual Review Of Environment And Resources **30**: 1-37.

1. Introduction

Boldo, E., S. Medina, et al. (2006). "Apheis: Health impact assessment of long-term exposure to PM2.5 in 23 European cities." European Journal Of Epidemiology **21**(6): 449-458.

Brimlecombe, P. (1987). "The big smoke." 165-169.

Bronnimann, S., E. Schuepbach, et al. (2000). "A climatology of regional background ozone at different elevations in Switzerland (1992-1998)." Atmospheric Environment **34**(29-30): 5191-5198.

BUWAL (1999). "Air Pollutant Emissions of Road Traffic 1950-2010 (Supplement)." Swiss Agency for the Environment, Forests and Landscape.

Fall, R. (2003). "Abundant oxygenates in the atmosphere: A biochemical perspective." Chemical Reviews **103**(12): 4941-4951.

Guenther, A., C. N. Hewitt, et al. (1995). "A Global-Model Of Natural Volatile Organic-Compound Emissions." Journal Of Geophysical Research-Atmospheres **100**(D5): 8873-8892.

Guttikunda, S. K., Y. H. Tang, et al. (2005). "Impacts of Asian megacity emissions on regional air quality during spring 2001." Journal Of Geophysical Research-Atmospheres **110**(D20).

Heck, W. W., W. W. Cure, et al. (1984). "Assessing Impacts Of Ozone On Agricultural Crops.1. Overview." Journal Of The Air Pollution Control Association **34**(7): 729-735.

Im, U., M. Tayanc, et al. (2006). "Analysis of major photochemical pollutants with meteorological factors for high ozone days in Istanbul, Turkey." Water Air And Soil Pollution **175**(1-4): 335-359.

Jorquera, H., R. Perez, et al. (1998). "Forecasting ozone daily maximum levels at Santiago, Chile." Atmospheric Environment **32**(20): 3415-3424.

Naja, M., H. Akimoto, et al. (2003). "Ozone in background and photochemically aged air over central Europe: Analysis of long-term ozonesonde data from Hohenpeissenberg and Payerne." Journal Of Geophysical Research-Atmospheres **108**(D2).

Nemmar, A., P. H. M. Hoet, et al. (2002). "Passage of inhaled particles into the blood circulation in humans." Circulation **105**(4): 411-414.

Ordonez, C., H. Mathis, et al. (2005). "Changes of daily surface ozone maxima in Switzerland in all seasons from 1992 to 2002 and discussion of summer 2003." Atmospheric Chemistry And Physics **5**: 1187-1203.

Poschl, U. (2005). "Atmospheric aerosols: Composition, transformation, climate and health effects." Angewandte Chemie-International Edition **44**(46): 7520-7540.

Riemer, D., W. Pos, et al. (1998). "Observations of nonmethane hydrocarbons and oxygenated volatile organic compounds at a rural site in the southeastern United States." Journal Of Geophysical Research-Atmospheres **103**(D21): 28111-28128.

Schade, G. W. and A. H. Goldstein (2006). "Seasonal measurements of acetone and methanol: Abundances and implications for atmospheric budgets." Global Biogeochemical Cycles **20**(1).

1. Introduction

Seinfeld, J. H. P., S.N. (1998). "Atmospheric chemistry and physics - from airpollution to climate change."

Sillman, S. (1999). "The relation between ozone, NOx and hydrocarbons in urban and polluted rural environments." Atmospheric Environment **33**(12): 1821-1845.

Singh, H. B., L. J. Salas, et al. (2004). "Analysis of the atmospheric distribution, sources, and sinks of oxygenated volatile organic chemicals based on measurements over the Pacific during TRACE-P." Journal of Geophysical Research-Atmospheres **109**(D15): art. no.-D15S07.

Stohl, A. T., T. (2001). "Experimental evidence for trans-Atlantic transport of air pollution." IGActivities Newsletter(No. 24): 10-12.

Tidblad, J., A. A. Mikhailov, et al. (2004). "Improved prediction of ozone levels in urban and rural atmospheres." Protection Of Metals **40**(1): 67-76.

2 The role of organic gases in tropospheric chemistry

Most of the VOCs in the troposphere are removed by oxidation, and this process starts with an initial reaction with either the hydroxyl radical (OH), the nitrate radical (NO_3), ozone (O_3) or photolysis. The peroxyradicals (RO_2), which are dominantly formed after OH or NO_3 attack, can react with NO and produce NO_2. This decreases the NO/NO_2 ratio, which is directly connected to the ozone budget (see section 2.2). Hence, in the presence of NO the photooxidation of VOCs leads to an elevated ozone mixing ratio.

This section introduces the basic oxidation mechanisms for the most important NMHCs and OVOCs (Section 2.1), their influence on the ozone budget and their tropospheric lifetimes (Section 2.2), section 2.3 summarizes the present knowledge of their emission sources and section 2.4 gives a brief overview of night-time chemistry.

2.1 Tropospheric photochemistry

2.1.1 Sources of hydroxyl radicals

The most important oxidant for the VOCs in the atmosphere is the OH radical. It is mostly formed by the photolysis of ozone. The sunlight which penetrates to the troposphere consists almost entirely of wavelengths longer than ca. 300 nm. The photolysis of ozone which occurs at wavelengths up to approx. 325 nm leads to electronically excited singlet oxygen ($O(^1D)$) radicals (equation 2-1). The main production of OH radicals follows from the reaction of the singlet oxygen radical with water vapour (equation 2-2). In rural regions of the UK photolysis of formaldehyde (HCHO) (equation 2-3) was found to give an important contribution to OH production (equations 2-4 and 2-5), in urban regions even dominating this process when the O_3 was suppressed by reaction with NO (Jenkin and Clemitshaw 2000) (equation 2-16).

$O_3 + h\nu \rightarrow O(^1D) + O_2$ (2-1)

$O(^1D) + H_2O \rightarrow 2\ OH^\bullet$ (2-2)

$HCHO + h\nu \rightarrow H^\bullet + RCO^\bullet$ (2-3)

$H^\bullet + O_2 + M \rightarrow HO_2^\bullet + M$ (2-4)

$HO_2^\bullet + NO \rightarrow HO^\bullet + NO_2$ (2-5)

2. The role of organic gases in tropospheric chemistry

2.1.2 Mechanisms of VOC oxidation

VOCs are oxidized by different pathways depending on their chemical characteristics and on the availability of reactants for the initial reaction.

2.1.2.1 Alkanes

The initial reaction is the abstraction of an H-atom by the OH radical producing an alkyl radical ($R^\bullet$) (equation 2-6).

$$RH + OH^\bullet \rightarrow R^\bullet + H_2O \tag{2-6}$$

This reaction is the rate limiting step in the following reaction sequence. The production rate of the alkyl radicals depends on the C-H bond energy as well as on the stability of the alkyl radical, which explains why the rate increases from primary to tertiary alkyl radicals. The stability also increases with the number of carbon atoms in the rest R (Atkinson 1994; Atkinson et al. 2006)

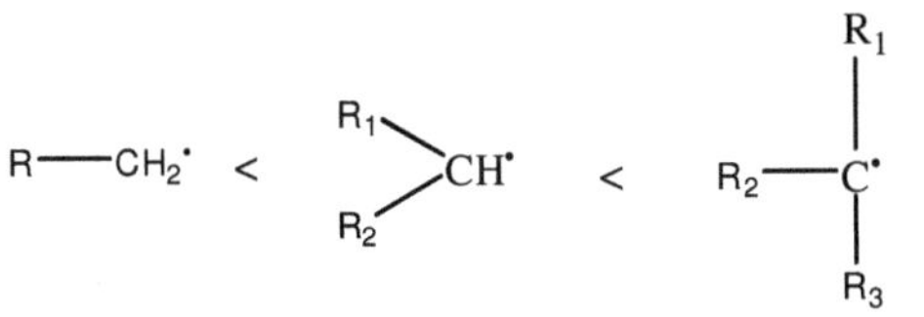

$$R-CH_2^\bullet \quad < \quad \begin{array}{c} R_1 \\ \searrow \\ CH^\bullet \\ \nearrow \\ R_2 \end{array} \quad < \quad R_2 - \underset{R_3}{\overset{R_1}{\underset{|}{\overset{|}{C^\bullet}}}}$$

Primary Secondary Tertiary

The alkyl radicals react fast with molecular oxygen in the presence of an air molecule (e.g. N_2) (equation 2-7). The resulting alkylperoxyradical ($RO_2^\bullet$) can react in several ways, but in anthropogenically polluted air the reaction with NO is the most common (equation 2-8).

$$R^\bullet + O_2 + M \rightarrow RO_2^\bullet + M \tag{2-7}$$

$$\begin{aligned} RO_2^\bullet + NO &\rightarrow RO^\bullet + NO_2 \\ &\rightarrow RONO_2 \end{aligned} \tag{2-8}$$

The products from this reaction are either alkoxyradicals ($RO^\bullet$) and NO_2 or alkylnitrates, whose gain increases with the length of the R-chain. Alkoxyradicals can react with molecular oxygen to produce aldehydes (equation 2-9).

$$RO^\bullet + O_2 \rightarrow R'CHO + HO_2^\bullet \tag{2-9}$$

2. The role of organic gases in tropospheric chemistry

They can also decompose or isomerize by a 1,5-H-Shift through a 6-membered transition state (only if R has more than 3 C-atoms). Figure 2-1 visualizes the different paths of reaction.

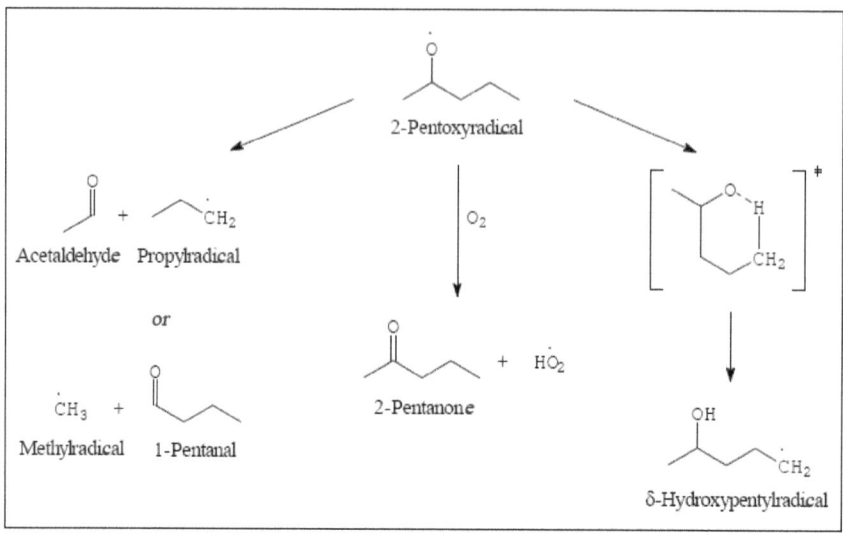

Figure 2-1. Pathways of decomposition of 2-Pentoxyradicals in ambient air (Winkler 2001)

The resulting alkyl- respectively hydroxyalkyl-radicals can react similar as in equation 2-7, the carbonyls react with OH-radicals or are photochemically decomposed, which again leads to the formation of OH- and HO_2-radicals. HO_2-radicals react with NO (equation 2-10) similar as in reaction 2-8.

$$HO_2^\bullet + NO \rightarrow OH^\bullet + NO_2 \qquad (2\text{-}10)$$

In this pathway of alkane decomposition, 2 NO molecules are transformed into NO_2 by the reactions 2-8 and 2-10, and by the decomposition of the resulting intermediates further NO is transformed into NO_2. As will be shown in section 2.2.1, the ratio NO_2/NO is decisive for the tropospheric ozone production.

2.1.2.2   Alkenes

The most common initial reaction pathway is the addition of an OH-radical to one of the >C=C< bonds to form β-hydroxyalkyl radicals (equations 2-11), as shown, for example, for 1-butene:

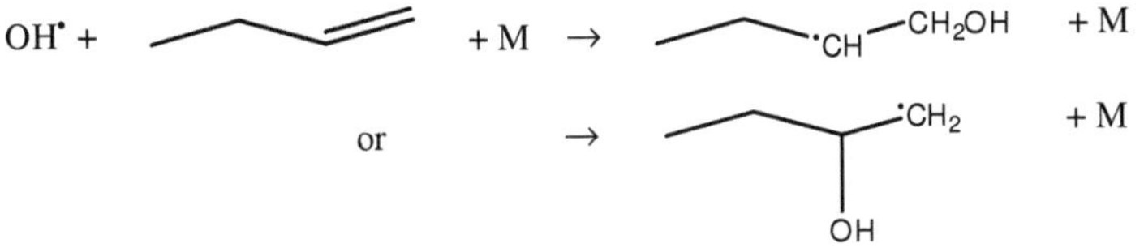

2. The role of organic gases in tropospheric chemistry

$$OH^\bullet + \text{(alkene)} + M \rightarrow \text{(·CH–CH}_2\text{OH)} + M \quad (2\text{-}11a)$$

or

$$\rightarrow \text{(·CH}_2\text{, OH)} + M \quad (2\text{-}11b)$$

Much less common is the H-atom abstraction from the C-H bonds of the alkyl substituent around the >C=C< bond (Atkinson 1994). As with the alkoxy radical in equation 2-7, the β-hydroxyalkyl radical react rapidly and solely with O_2 (equation 2-12) to form β-hydroxyalkyl peroxy radicals.

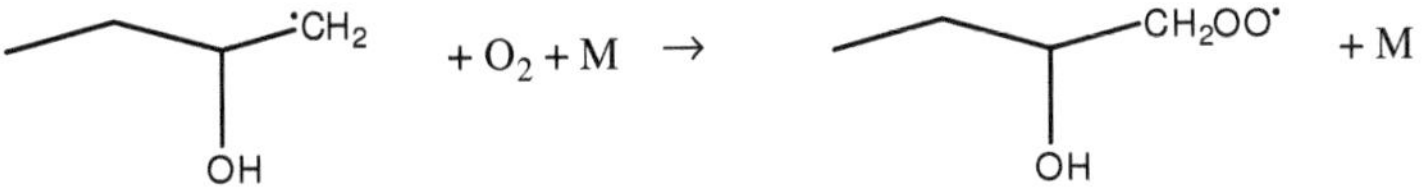

$$\text{(·CH}_2\text{, OH)} + O_2 + M \rightarrow \text{(CH}_2OO^\bullet\text{, OH)} + M \quad (2\text{-}12)$$

This product reacts in an analogous way as in equation 2-8 with NO to form b-hydroxyalkoxy radicals, which either decomposes, reacts with O_2 to form aldehydes or isomerise by 1,5-H-shift through a 6-membered transition state.

Due to their higher reactivity, alkenes tend to react readily with ozone and NO_3. The reaction with NO_3 is estimated to explain about 30 % of the alkene oxidation on a daily basis (Geyer et al. 2001). The reaction of alkenes with O_3 occurs by an electrophilic addition to the double bond creating an excited ozonide, which decomposes into a carbonyl and Crigee-radical (see figure 2-2).

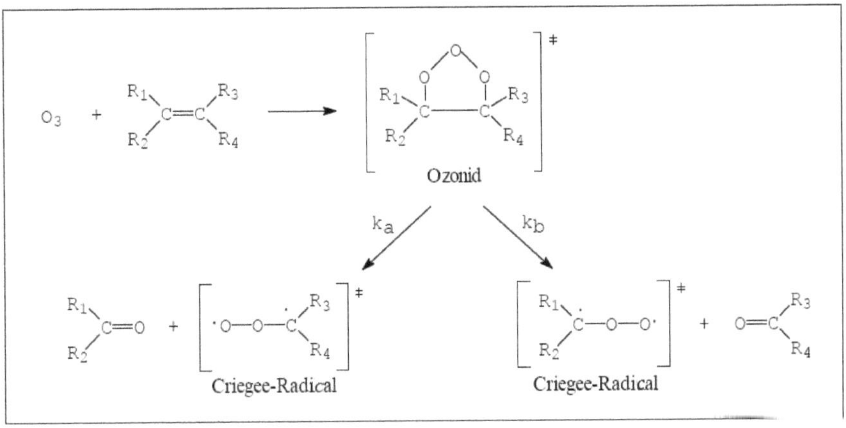

Figure 2-2. Reactions of alkenes with ozone (Winkler 2001).

2. The role of organic gases in tropospheric chemistry

The Criegee-radical is unstable and either decomposes or reacts further to produce carbonyls, organic acids, esters and hydroperoxides (Atkinson 1997).

2.1.2.3   Alkynes

Due to the lower reactivity of the triple bond, the decomposition of the alkynes is much slower (Atkinson 2006) than for other NMHCs. The initial step is in this case also the addition of an OH-radical to the triple bond.

2.1.2.4   Aromatics

The aromatic compounds react with the OH and the NO_3 radical, for which the reaction with the OH radical dominates (Atkinson 2000). This reaction proceeds either by H-atom subtraction from the C-H bonds of the alkyl substituent groups (or in the case of benzene from the aromatic ring) (R1), or by OH radical addition to the aromatic ring (R2). In R1 the stable product is a carbonyl compound (e.g. transformation of toluene to benzaldehyde). R2 is the primary pathway for decomposition of aromatic hydrocarbons. After the addition of the OH radical to the aromatic ring, a cyclic hydroxyl peroxy radical is produced from the reaction with molecular oxygen. The ring structure opens in the following step and formation of epoxy compounds, different saturated and unsaturated dicarbonyl radicals and finally methyl glyoxal has been observed. Due to a large number of laboratory studies on the atmospheric oxidation of aromatic hydrocarbons in the last year (see the extensive review by Calvert et al. (2000)), the understanding of the detailed chemistry involved has been improved considerably. However, ring opening pathways (R2) are still speculative, whereas ring-retaining pathways (R1) are comparably well understood (Hamilton and Lewis 2003).

2.1.3   Destruction mechanisms for the OVOCs

OVOCs include a variety of compounds which are decomposed differently in ambient air. Carbonyls either react with the OH radical or by photolysis. However, most carbonyls predominantly decompose by the OH-reaction and only in. the case of formaldehyde and some dicarbonyls the photolysis becomes the dominant initial reaction (equations 2-3 and 2-13, 2-14, 2-4):

$HCHO + hv \rightarrow CO + H_2$ (2-13)

$HCHO + hv \rightarrow HCO + H$ (2-3)

$HCO + O_2 \rightarrow HO_2 + CO$ (2-14)

$H + O_2 + M \rightarrow HO_2 + M$ (2-4)

2. The role of organic gases in tropospheric chemistry

The initial degradation steps for aldehydes by OH radicals followed by O_2 oxidation (equation 2-15) proceed as shown in the example for acetaldehyde:

$$CH_3CHO + OH + O_2 \rightarrow \rightarrow CH_3C(O)O_2^\bullet \qquad (2\text{-}15)$$

The produced peroxy radical is a precursor for the PAN formation, if enough NO_2 is available. In the presence of sufficient NO, however, the peroxy radical of a C_n-aldehyde reacts dominantly with NO, leading to a C_{n-1}-aldehyde and CO_2. For the OH-radical initiated degradation of ketones the reaction proceeds by H-atom abstraction and subsequent formation of alkoxy radicals.

Alcohols and ethers are removed by the OH and the NO_3 radical reactions. The OH reaction, being the main path of removal, proceeds in the case of alcohols by H-atom abstraction from the various C-H bonds and the O-H bond. The H-atom abstraction occurs in the case of ethers mainly from the C-H bonds on the carbon atoms adjacent to the ether -O- atom (Atkinson 2000). The esters are solely removed by the reaction with the OH radical.

2.2 Daytime photo oxidant formation

2.2.1 Photostationary state

In the stratosphere ozone production is initiated by the photo dissociation of molecular oxygen at wavelengths $\lambda \leq 242$ nm (Chapman, 1930). As this shortwave radiation is absorbed in the stratosphere, no ozone is produced in the troposphere by this reaction. It was only after the discovery of photochemical smog in Los Angeles in the 1950s that the mechanisms leading to the production of tropospheric ozone were discovered (Haagen-Smit 1952). The crucial initiating step is the photolysis of nitrogen dioxide (NO_2) to nitrogen monoxide (NO), which only requires solar radiation with wavelengths $\lambda \leq 410$ nm. The following reaction of the produced atomic oxygen ($O(^3P)$) with molecular oxygen is fast and produces ozone in the presence of an inert collision partner M like N_2 (equations 2-16 to 2-17).

$$NO_2 + h\nu\ (\lambda \leq 410\ nm) \xrightarrow{j_{1.1}} NO + O(^3P) \qquad (2\text{-}16)$$

$$O_2 + O + M \xrightarrow{k_{1.2}} O_3 + M \qquad (2\text{-}17)$$

However, ozone reacts rapidly with NO leading to ozone destruction (equation 2-18):

$$O_3 + NO \xrightarrow{k_{1.3}} O_2 + NO_2 \qquad (2\text{-}18)$$

2. The role of organic gases in tropospheric chemistry

The combination of these three reactions results in a zero cycle, leading to no net ozone production. Reaction 2-17 is very fast leading to the immediate reaction of the atomic oxygen with O_2, so the rate determining step for this cycle is equation 2-16. Assuming steady state conditions (ss) the ozone concentration can be calculated by the equation 2-19:

$$[O_3]_{ss} = \frac{j_{1.1}[NO_2]}{k_{1.3}[NO]} \tag{2-19}$$

2.2.2 Ozone formation in polluted environments

Following section 2.2.1 no net ozone formation occurs. The high observed ozone concentrations in urban and rural regions therefore have to be caused by additional reactions, and this is where the VOCs play an important role in "fuelling" the ozone production. In order to explain ozone formation, some other species must add atomic oxygen to the reaction system of reactions 2-16 to 2-18. Hydroperoxy (HO_2) and organic peroxy radicals (RO_2) have this ability, and they are produced as intermediates in the photochemical oxidation of VOCs as shown in section 2.1 and of carbon monoxide (CO). Figure 2-3 shows the oxidation cycle for alkanes, which produces both ozone and aldehydes.

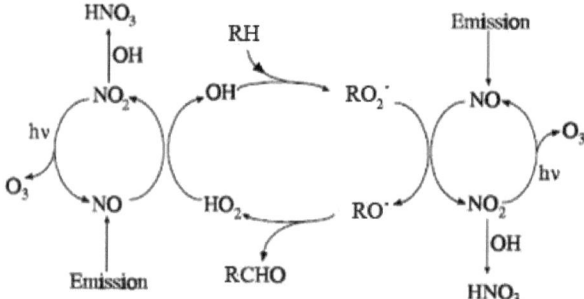

Figure 2-3. The oxidation cycle for alkanes and tropospheric ozone formation.

In urban polluted regions NO mixing ratios are large enough to produce ozone according to Figure 2-3. Figure 2-4 shows the ozone production as a function of NO_x for various levels of VOCs. Under high NO_x conditions, the ozone production increases with decreasing NO_x for a given level of VOCs. At a certain point, the ozone production reaches its maximum as the chemistry undergoes a transition state from high-NO_x to low-NO_x conditions. Thereafter, the ozone production decreases rapidly as the amount of NO_x approaching a level only found under clean air conditions. At higher NO_x mixing ratios, the ozone production increases with higher VOC mixing ratios. The transition state is associated with a specific VOC to NO_x ratio which also depends on the mixing ratios of single

2. The role of organic gases in tropospheric chemistry

VOC compounds and their reactivities. In Switzerland, the large reductions in VOC and NO_x emissions caused by the introduction of catalytic converters in gasoline driven vehicles and emission reductions from the industry, led to decreasing ozone maxima at few selected sites since the beginning of the 1990's (Ordonez et al. 2005). However, the number of days with ozone levels above the set limit has not significantly changed in this time period.

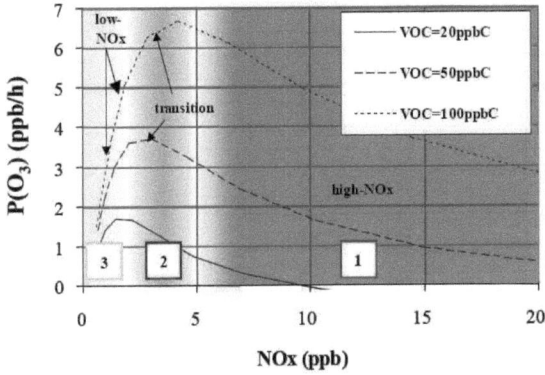

Figure 2-4. Net rates of ozone production (ppb per hour) as a function of NOx (ppb), for VOC = 20 ppbC (solid line), 50 ppbC (long dashed line) and 100 ppbC (short dashed line) (Spirig 2003).

2.2.3 Photochemical Ozone Creation Potential (POCP) and tropospheric lifetimes

An important scale for quantifying the produced ozone from the oxidation of specific VOCs under European conditions was introduced by Derwent et al. (1998). It is derived from numerical simulations based on a detailed chemical reaction mechanism for atmospheric conditions (Master Chemical Model, MCM), which is connected with a trajectory model. This model calculates the incremental ozone production from a parcel of air to which an additional emission of 4,7 kg/km² of each VOC_i is added. The $POCP_i$ value is given by the increase in the ozone mixing ratio caused by VOC_i, and is always given as the relative values compared the POCP value of ethene ($POCP_{ethene}$=100). In Table 2 the POCP values for the compounds measured in this study are listed. Not all values were available from the last study applying MCM version 3 (Saunders et al. 2003), the POCP values marked with * are from the earlier study by Derwent el al. (1998), and a few POCP values (with **) was from another study by Altenstedt et al. (2000).

2. The role of organic gases in tropospheric chemistry

Compounds	POCP value		Lifetimes due to Reaction with ˙OH^a	Reference
Formaldehyde	47.1		1.2 day	1)
Acetaldehyde	55		0.7 day	1)
Propanal	61.2		14.2 hr	1)
Butanal	58.5		11.8 hr	1)
Pentanal	59.7		9.7 hr	1)
Hexanal	na		8.8 hr	4)
Benzaldehyde	-9.2	*	1.0 day	5)
Acrolein	73-126	**	14.0 hr	1)
Methacrolein	95-110	**	3.5 day	1)
MTBE	17.8		3.9 day	1)
Acetone	7.5		52.8 day	1)
MVK	na		14.8 hr	1)
MEK	35.3		10.1 day	1)
Methanol	16.5		12.3 day	1)
Ethanol	39.7		3.5 day	1)
Iso-propanol	19.8		2.2 day	1)
n-Propanol	53.1		2.1 day	1)
2-Methyl-3-buten-2-ol	na		4.4 hr	3)
n-Butanol	53.9		1.4 day	1)
Methyl acetate	5.1		33.9 day	2)
Ethyl acetate	23.8		7.7 day	2)
Butyl acetate	28.9		2.8 day	2)
Butane	36.3		4.6 day	1)
1,3-Butadiene	76		4.2 hr	1)
Isoprene	101.6		2.8 hr	1)
Benzene	21.8	*	9.4 day	1)
Toluene	63.7	*	1.9 day	1)
Ethylbenzene	73	*	1.6 day	1)
m,p-Xylene	105.9	*	14.7 hr	1)
o-Xylene	105.3	*	20.3 hr	1)
1,2,4-Trimethylbenzene	127.8	*	8.5 hr	1)

Table 2-1. Measured compounds with their values for Photochemical Ozone Creation Potentials (POCPs) and tropospheric lifetimes with the reaction with the OH radical. References: 1) (Atkinson 1994), 2) (Wallington et al. 1988), 3) (Rudich et al. 1995), 4) (D'Anna et al. 2001), 5) (Semadeni et al. 1995).

^{a} For a 12 hr daytime average ˙OH radical concentration of 1.0 x 10^6 molecules cm^{-3} and 298 K.

2.3   Sources of OVOCs

The relative abundance of OVOCs in ambient air strongly depends on the measurement site. Their contributions to ambient air are estimated to be in the order of 20 – 50 % of the total content of organic compounds in air in industrialized countries (Ciccioli et al. 1994; Goldan et al. 1995; Bundesamt für Umwelt 2000). Furthermore, OVOCs have even been identified in air samples from remote sites such as Antarctica (Boudries et al. 2002), which confirms their ubiquitous occurrence in the troposphere. There are three possible sources for OVOCs: emissions from the biosphere, anthropogenic emissions and oxidation of VOCs in the atmosphere. Table 2-2 summarizes the known sources for VOCs quantified in this thesis.

2. The role of organic gases in tropospheric chemistry

The biosphere and higher plants in particular emit large quantities of reactive VOCs into the atmosphere. The number of different compounds is estimated to be in the order of several thousands, though only few are emitted at rates that significantly influence the chemical composition of the atmosphere. Isoprene and the monoterpenes are the most abundant compounds emitted from these sources and due to their reactivity they are oxidized rapidly in the atmosphere producing numerous oxygenated compounds. A number of oxygenated compounds are also emitted directly by biogenic sources. Oxygenated C_1-C_3 VOCs are found to be the dominant VOC components in forest air (Goldan et al. 1995), and the most abundant OVOCs were methanol, acetone and acetaldehyde. The biogenic production processes of these compounds are still poorly understood. Methanol is emitted by a variety of plants and continuous emissions have been observed in many ecosystems, including forest and grasslands. Methanol and acetone are frequent intermediates of the plant metabolism and are found in many tissues of the plants (Fall 2003). Ethanol and acetaldehyde are known to occur during fermentation processes. However, biogenic emissions of ethanol and acetaldehyde have also been observed in fairly dry regions where fermentation is unlikely to be of importance, suggesting other, yet unknown processes.

Anthropogenic sources of OVOCs are either due to combustion processes or solvent usage, from which combustion is the most dominant (Friedrich 1999). Formaldehyde has been the most abundant carbonyl measured in many urban studies (Possanzini et al. 1996; Granby et al. 1997; Sin et al. 2001), whereas at a road site in Stockholm also large mixing ratios of methanol (up to 72 ppb), ethanol (up to 247 ppb) and acetone (up to 129 ppb) were found during winter (Jonsson et al. 1985). Methyl-tert-butyl-ether (MTBE) is used as a fuel additive and emitted by both vehicle exhaust during cold starts and from fuel evaporation (Poulopoulos and Philippopoulos 2000; Kawamoto et al. 2003). This compound is toxic and water soluble and could therefore be a threat to drinking water. Ethers, alcohols, acetates and ketones are often applied as replacement products for CFCs. For instance CFC-113, which was used intensively for cleaning electronic components, is now replaced by liquids containing iso-propanol (Isakson 1994).

The oxidation of atmospheric NMHCs is also a large source of OVOCs. They are formed by a formed by a variety of reactions (section 2.1), mostly initiated by reaction with the OH radical. The details of the reaction pathways following this step are dependent on the amount of NO and NO_2 available, even if other compounds also have an influence (see Figure 2-3). Formaldehyde is one of the main products from the oxidation of numerous VOCs (Atkinson 2000). Holzinger et al. (2005) reported large production of acetone and methanol from aged biomass burning plumes. In Pittsburgh secondary production was reported to account for 12 - 27% of the total mixing ratios for

2. The role of organic gases in tropospheric chemistry

acetaldehyde, butanone and acetone in winter and 26 - 34% of their mixing ratio in summer (Millet et al. 2005).

Compounds	Identified sources				
	Biogenic sources	Anthropogenic sources			Photooxidation
		Road traffic	Stationary combustion	Solvent	
Formaldehyde		X	X		X
Acetaldehyde	X	X	X		X
Propanal		X	X		X
Butanal		X	X		X
Pentanal		X	X		X
Hexanal	X	X	X		X
Benzaldehyde		X	X		X
Acrolein		X	X		X
Methacrolein		X	X		X
Methyl-tert-butyl-ether		X			
Acetone	X	X	X		X
MVK		X	X		X
MEK	X	X	X		X
Methanol	X	X	X	X	X
Ethanol	X	X	X	X	
Iso-propanol		X	X	X	
n-Propanol				X	
2-Methyl-3-buten-2-ol	X	X			
n-Butanol				X	
Methyl acetate		X	X		
Ethyl acetate		X	X	X	
Butyl acetate				X	
Butane		X	X		
1,3-Butadiene		X	X		
Isoprene	X	X	X		
Benzene		X	X	X	
Toluene		X	X	X	
Ethylbenzene		X	X		
m,p-Xylene		X	X	X	
o-Xylene		X	X	X	
1,2,4-Trimethylbenzene		X	X		

Table 2-2. Measured compounds listed with their sources known from the literature.

2.4 Night-time oxidation

This section explains shortly the most important features of the nigh-time oxidation of VOCs including the mechanism of the oxidation of NMHCs and OVOCs with NO_3. During the night OH radicals are not produced by photochemistry as shown in section 2.2.2. The NMHCs and OVOCs react either with NO_3 or O_3, for which only unsaturated organic compounds react with O_3. However, NO_2 can react with O_3 to produce NO_3:

$$NO_2 + O_3 \rightarrow NO_3 + O_2 \qquad (2\text{-}20)$$

NO_3 reacts fast with NO and is photolyzed immediately at daytime. This compound has therefore no significant impact on the tropospheric chemistry after sunrise. In the night though, NO_3 can act as a strong oxidation agent. The reaction with NO_3 is for several compounds a significant decomposition pathway. It reacts with alkanes and aldehydes by H-atom abstraction (equation 2-21 and 2-22) producing alkyl- and alkoxy- radicals and nitric acid (HNO_3):

2. The role of organic gases in tropospheric chemistry

$$RH + NO_3 \rightarrow R^{\bullet} + HNO_3 \quad (2\text{-}21)$$

$$RCHO + NO_3 \rightarrow {}^{\bullet}RCO + HNO_3 \quad (2\text{-}22)$$

HNO_3 is water soluble and is therefore removed from the air mostly by wet deposition. The alkyl- and alkoxy- radicals react further as shown in equation 2-7 and 2-9. Night-time oxidations are much slower than during the day and still a topic of recent research (McLaren et al. 2004).

References:

Altenstedt, J. and K. Pleijel (2000). "An alternative approach to photochemical ozone creation potentials applied under European conditions." Journal Of The Air & Waste Management Association **50**(6): 1023-1036.

Atkinson, R. (1994). "Gas-Phase Tropospheric Chemistry Of Organic-Compounds." Journal Of Physical And Chemical Reference Data: R1-&.

Atkinson, R. (1994). "Gas-phase tropospheric chemistry of organic compounds." J. Phys. Chem. Ref. Data **2**: 1-216.

Atkinson, R. (1997). "Gas-phase tropospheric chemistry of volatile organic compounds.1. Alkanes and alkenes." Journal Of Physical And Chemical Reference Data **26**(2): 215-290.

Atkinson, R. (2000). "Atmospheric chemistry of VOCs and NOx." Atmospheric Environment **34**(12-14): 2063-2101.

Atkinson, R., D. L. Baulch, et al. (2006). "Evaluated kinetic and photochemical data for atmospheric chemistry: Volume II - gas phase reactions of organic species." Atmospheric Chemistry And Physics **6**: 3625-4055.

Boudries, H., J. W. Bottenheim, et al. (2002). "Distribution and trends of oxygenated hydrocarbons in the high Arctic derived from measurements in the atmospheric boundary layer and interstitial snow air during the ALERT2000 field campaign." Atmospheric Environment **36**(15-16): 2573-2583.

Bundesamt für Umwelt, W. u. L., BUWAL (2000). "Immisionsmessungen polarer VOC in der Schweiz 1999."

Calvert, J. G. A., R.; Kerr, J.A.;Madronich, S.;Moorgat, G.;Wallington, T.;Yarwool, G. (2000). "The Mechanisms of Atmospheric Oxidation of the Alkenes."

Ciccioli, P., A. Cecinato, et al. (1994). "Polar Volatile Organic-Compounds (Voc) Of Natural Origin As Precursors Of Ozone." Environmental Monitoring And Assessment **31**(1-2): 211-217.

D'Anna, B., W. Andresen, et al. (2001). "Kinetic study of OH and NO3 radical reactions with 14 aliphatic aldehydes." Physical Chemistry Chemical Physics **3**(15): 3057-3063.

2. The role of organic gases in tropospheric chemistry

Derwent, R. G., M. E. Jenkin, et al. (1998). "Photochemical ozone creation potentials for organic compounds in northwest Europe calculated with a master chemical mechanism." Atmospheric Environment **32**(14-15): 2429-2441.

Fall, R. (2003). "Abundant oxygenates in the atmosphere: A biochemical perspective." Chemical Reviews **103**(12): 4941-4951.

Friedrich, R. O., A. (1999). "Anthropogenic Emissions of volatile organic compounds." in: Reactive Hydrocarbons in the Atmosphere, edited by Hewitt, C. N.: 1-39.

Geyer, A., B. Alicke, et al. (2001). "Chemistry and oxidation capacity of the nitrate radical in the continental boundary layer near Berlin." Journal Of Geophysical Research-Atmospheres **106**(D8): 8013-8025.

Goldan, P. D., W. C. Kuster, et al. (1995). "Hydrocarbon measurements in the southeastern United States: The Rural Oxidants in the Southern Environment (ROSE) program 1990." Journal Of Geophysical Research-Atmospheres **100**(D12): 25945-25963.

Goldan, P. D., M. Trainer, et al. (1995). "Measurements Of Hydrocarbons, Oxygenated Hydrocarbons, Carbon-Monoxide, And Nitrogen-Oxides In An Urban Basin In Colorado - Implications For Emission Inventories." Journal Of Geophysical Research-Atmospheres **100**(D11): 22771-22783.

Granby, K., C. S. Christensen, et al. (1997). "Urban and semi-rural observations of carboxylic acids and carbonyls." Atmospheric Environment **31**(10): 1403-1415.

Haagen-Smit, A. J. (1952). "Chemistry and physiology of Los Angeles smog." Ind. Eng. Chem. **44**: 1342-1346.

Hamilton, J. F. and A. C. Lewis (2003). "Monoaromatic complexity in urban air and gasoline assessed using comprehensive GC and fast GC-TOF/MS." Atmospheric Environment **37**(5): 589-602.

Holzinger, R., J. Williams, et al. (2005). "Oxygenated compounds in aged biomass burning plumes over the Eastern Mediterranean: evidence for strong secondary production of methanol and acetone." Atmospheric Chemistry And Physics **5**: 39-46.

Isakson, K. V., A.L (1994). "An assessment of alternatives for replacing Freon 113 in bench type electrical circuit board cleaning at Fermi National Accelerator Laboratory."

Jenkin, M. E. and K. C. Clemitshaw (2000). "Ozone and other secondary photochemical pollutants: chemical processes governing their formation in the planetary boundary layer." Atmospheric Environment **34**(16): 2499-2527.

Jonsson, A., K. A. Persson, et al. (1985). "Measurements Of Some Low Molecular-Weight Oxygenated, Aromatic, And Chlorinated Hydrocarbons In Ambient Air And In Vehicle Emissions." Environment International **11**(2-4): 383-392.

Kawamoto, K., J. S. Arey, et al. (2003). "Emission and fate assessment of methyl tertiary butyl ether in the Boston area airshed using a simple multimedia box model: Comparison with urban air measurements." Journal Of The Air & Waste Management Association **53**(12): 1426-1435.

2. The role of organic gases in tropospheric chemistry

McLaren, R., R. A. Salmon, et al. (2004). "Nighttime chemistry at a rural site in the Lower Fraser Valley." Atmospheric Environment **38**(34): 5837-5848.

Millet, D. B., N. M. Donahue, et al. (2005). "Atmospheric volatile organic compound measurements during the Pittsburgh Air Quality Study: Results, interpretation, and quantification of primary and secondary contributions." Journal Of Geophysical Research-Atmospheres **110**(D7).

Ordonez, C., H. Mathis, et al. (2005). "Changes of daily surface ozone maxima in Switzerland in all seasons from 1992 to 2002 and discussion of summer 2003." Atmospheric Chemistry And Physics **5**: 1187-1203.

Possanzini, M., V. Dipalo, et al. (1996). "Measurements of lower carbonyls in Rome ambient air." Atmospheric Environment **30**(22): 3757-3764.

Poulopoulos, S. and C. Philippopoulos (2000). "Influence of MTBE addition into gasoline on automotive exhaust emissions." Atmospheric Environment **34**(28): 4781-4786.

Rudich, Y., R. Talukdar, et al. (1995). "Reaction Of Methylbutenol With Hydroxyl Radical - Mechanism And Atmospheric Implications." Journal Of Physical Chemistry **99**(32): 12188-12194.

Saunders, S. M., M. E. Jenkin, et al. (2003). "Protocol for the development of the Master Chemical Mechanism, MCM v3 (Part A): tropospheric degradation of non-aromatic volatile organic compounds." Atmospheric Chemistry And Physics **3**: 161-180.

Semadeni, M., D. W. Stocker, et al. (1995). "The Temperature-Dependence Of The Oh Radical Reactions With Some Aromatic-Compounds Under Simulated Tropospheric Conditions." International Journal Of Chemical Kinetics **27**(3): 287-304.

Sin, D. W. M., Y. C. Wong, et al. (2001). "Trends of ambient carbonyl compounds in the urban environment of Hong Kong." Atmospheric Environment **35**(34): 5961-5969.

Spirig, C. (2003). "Biogenic volatile organic compounds and their role in the formation of ozone and aerosols." Doctoral Thesis ETH No. 15557.

Wallington, T. J., P. Dagaut, et al. (1988). "The Gas-Phase Reactions Of Hydroxyl Radicals With A Series Of Esters Over The Temperature-Range 240-440-K." International Journal Of Chemical Kinetics **20**(2): 177-186.

Winkler, J. (2001). "Atmosphärische Nichtmethan-Kohlenwasserstoffe in Berlin/Brandenburg: Messtechnik, Bestandsaufnahme und Beiträge zur lokalen Photoooxidatienbildung." Johann Wolfgang Goethe-University: 12-18.

3. Measurement of organic gases in ambient air and instrument development

3.1 Introduction

Due to the diversity of the organic compounds and their low abundances in the atmosphere several techniques have been developed to identify and characterize these compounds. The most commonly used instrument is still the capillary gas chromatograph (GC), which has been applied for several decades (Ioffe et al. 1977; Helmig 1999). For polar and reactive compounds, other techniques involving derivatization have also been developed. The most frequently used sampling method uses 2,4-dinitrophenylhydrazine (DNPH) coated C-18 cartridges, which were developed for analyzing ketones and aldehydes. After extraction with acetonitrile, the extract is analyzed by High Performance Liquid Chromatography (HPLC) with UV detection (Cao 1999). The main disadvantage of this system is the amount of user interaction needed. In recent years direct introduction mass spectrometry has been developed to analyze several compounds, even reactive, at high frequency. Proton transfer reaction-mass spectrometry (PTR-MS) and chemical ionization reaction-mass spectrometry (CIR-MS) are newly developed techniques which have proven their usability in several studies (Wisthaler et al. 2002; Ammann et al. 2004; Steinbacher 2004; Blake et al. 2006). These systems deliver high frequency data, but have limited sensitivity and cannot separate compounds with the same molecular mass. Differential Optical Absorption Spectrometry (DOAS) allows simultaneous detection of formaldehyde, some mono-aromatic hydrocarbons and inorganic gases (SO_2, NO_2, O_3 and HNO_2). It has shown excellent agreement with conventional methods for formaldehyde and the inorganic gases. But low correlation with conventional methods raise questions about its reliability for analyzing the mono-aromatic compounds (Cao 1999).

3.2 GC systems for OVOC analysis

The main strength of GC analysis is the ability to separate almost all compounds by their different chemical and physical properties. Due to the ultra-low levels of the compounds in air, they have to be pre-concentrated prior to analysis. This can either be performed in a cryogenic trap, on an adsorbent material or a combination of both. Sampling of OVOCs for GC analysis has been performed on various adsorbent materials. Both passive sampling, where the transport of the analytes to the sampling medium is caused by diffusion, and active sampling, where the analytes are drawn through an adsorbent tube packed with one or more adsorbents, have been applied in the OVOC analysis. The active sampling is predominantly used, and desorption from the adsorbent is mostly performed by heating. The adsorbent materials used are activated carbon, carbon molecular sieves, porous organic polymers and graphitized carbon blacks (Dettmer and Engewald 2003). One well known problem with adsorbent sampling is the artefact production by the reaction of ozone with the adsorbent material. Several techniques have been applied for ozone removal (Helmig 1997), but they all have their limitations and should only be used after careful validation. Another problem of the adsorbent sampling is the water management. The humidity in ambient air reduces the capacity of adsorbents and causes problems with blocking of sub-zero cooled adsorbent traps. Small amounts of water reduce the lifetime of the chromatographic column as well. To ensure good separation of the OVOCs a semi-polar to polar stationary phase is needed, which is relatively water sensitive. Water removal techniques used are traps with drying agents, ion exchange membranes (Nafion), freeze out traps, and dry purging of adsorbent after sampling (Helmig and Vierling 1995).

3.3 Instrument development

In this section the instrument development, which was performed in close cooperation with the University of Bristol, is presented. The system was based on an adsorption desorption system (ADS) coupled to a gas chromatograph-mass spectrometer (GC-MS) for analysis of halocarbons (Simmonds et al. 1995). For adapting this type of instrument to OVOC analysis a number of changes had to be made. Water management, ozone removal, calibration, separation and inertness of the system were issues that had to be solved. In Figure 3-1 a schematic overview over the Modified ADS (MADS) for OVOC analysis is shown. In short, the analytes were sampled on a 0.6 g Hayesep-D bulk trap at room temperature. By this procedure the majority of the water was not retained on the trap. The analytes were then released by heating the trap to 160 °C and were re-focused on a 14 mg Hayesep D trap (- 40 °C) from where they were transferred to the analysis system (GC-MS) by desorption at 180 °C.

The following sections present the test for the development of the system. The work was performed at Empa in Dübendorf, at the Atmospheric Chemistry group at the University of Bristol and at the laboratory for atmospheric chemistry at Paul Scherrer Institute (PSI) in Villigen, Switzerland. The gas standards used in these tests were prepared by Apel & Riemer Environmental Inc., USA. In appendix II the different gas standard mixtures used in these tests and later in the measurement campaigns are summarized.

3. Measurement of organic gases in ambient air and instrument development

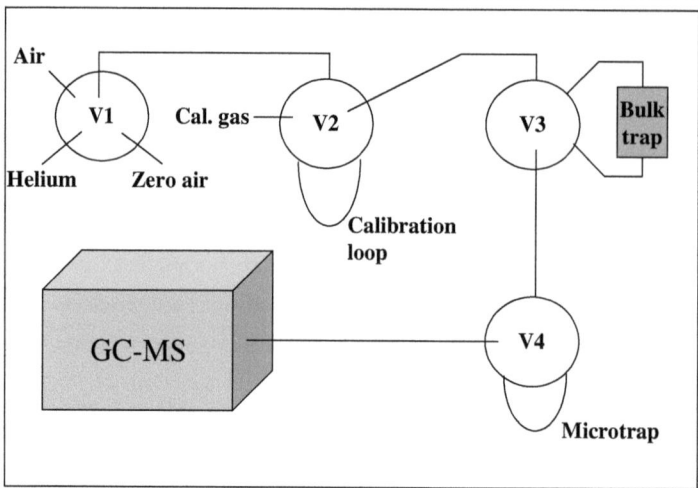

Figure 3-1. Schematic overview of the modified adsorption desorption system (MADS) coupled with GC-MS.

3.3.1 Tubing

Tests of different types of tubings for suitability for the analysis of OVOCs were performed at PSI on a PTR-MS. Tests were performed on 1 m and 10 meter long ¼" tubes of PFA, FEP, PTFE, stainless steel, sulfinated silcosteel (Restek) and deactivated fused silica (0.53 mm ID). The silcosteel line showed a peculiar feature with a strong hold-back of all compounds at the start of the experiment under dry conditions. After introducing humid air into the line, the properties were similar to PFA and deactivated fused silica (0.53 mm ID), the best materials in the test. For PFA and Silcosteel also 1/8" dimension was tested with similar results. Silcosteel tubing was additionally tested in the size 1/16" for use in the inner part of the instrument, and it proved to be as suitable as the other dimensions. For this reason also the particle filter was used in the same material, namely silcosteel (Restek) with a pore size of 0.5 µm.

3.3.2 Water removal

Various approaches for water removal for measurements of OVOCs are reported in the literature. Often cold traps are used, which consist of wide bore tubes, held at sub-zero temperatures to freeze the water to the walls. Apel et al. (2003) and Hopkins et al. (2003) used 1/8" silcosteel at -15 °C and a trap made of glass at -30 °C respectively. Adsorptive removal of water by sampling the compounds on a pre-column is another straight forward approach (Leibrock and Slemr 1997; Riemer et al. 1998). The analytes are refocused on a second trap before transfer to the analytical column.

Water can also be removed by crystalline salts as in the study of Monod et al. (2003), which used Chromosorb W AW coated with lithium chloride (LiCl) to remove water from the sample before the analysis of gas phase ethanol.

The following paragraphs present the tests which have been performed for water removal in order to find the best approach for the targeted range of OVOCs. Water removal by cold trap and the pre-column CP Wax 52 CB were tested at Empa. Tests were also performed with molecular sieve (3 Å), which removed water efficiently, but also removed large amounts of the OVOCs. Additionally, tests for the LiCl-method were performed by D. Young at the University of Bristol.

3.3.2.1 Cold trap

The cold trap was made by placing a U-shaped tube of either silcosteel or PFA in a Dewar with ethanol kept at -20 °C or -40 °C with a Neslab CC-65 cooler. Table 3-1 gives an overview of the tests performed on the cold traps. Homogeneous temperature conditions in the Dewar were obtained by using a lab mixer. All tubing except for the trap itself was made of 1/4" PFA. The sample air was produced by mixing humidified synthetic air with the OVOC standard mixture (Empa-standard 1 – see appendix II). Before connecting the cold trap, 3 samples were taken as reference in each experiment. Then the trap was connected, and after flushing for 5 minutes, the first sample was collected. In order to delay the creation of an ice plug, the flow was reduced to 10 ml/min between the samples. After readjusting the flow, the next sample was collected after 5 minutes. Total capacity of the 1/8" silcosteel cold trap was 4,3 litres of humid air (90% relative humidity at 23 °C). All samples were collected with a 0.5 ml gastight syringe and analyzed by GC-MS.

Test	% Apel standard in humid air	RH (%)	Temperature water trap	Trap length and material	Total flow (ml/min)	Concentration (ppb)
1	10	85%	-20°C	13 cm 1/8" Silcosteel	100	~40
2	10	85%	-40°C	13 cm 1/8" Silcosteel	100	~40
3	10	85%	-20°C	24 cm 1/4" Silcosteel	100	~40
4	10	85%	-40°C	24 cm 1/4" Silcosteel	100	~40
5	10	85%	-20°C	15 cm 1/8" PFA	100	~40
6	10	85%	-40°C	15 cm 1/8" PFA	100	~40

Table 3-1. Test performed for losses of OVOC in a cold trap.

The Figures 3-2 to 3-7 show the average values for the reference samples before and the values for the samples collected after the cold trap. Figure 3-2 and 3-3 show the results from 1/8" silcosteel. There were no losses to be observed at -20 °C, but since many compounds showed higher mixing ratios after the trap, which must be due to adsorption and release effects at the start of the experiment,

3. Measurement of organic gases in ambient air and instrument development

more experiments were needed. Ethanol is not shown for the first two tests due to contamination from coolant and the high scatter in the benzaldehyde mixing ratios was suspected to be due to adsorption effect on the metal surfaces in the pressure reduction valve. At -40 °C losses were observed for iso-propanol and higher boiling alcohols (see Figure 3-3), and the effect of the increased amount after trap was reduced. After increasing the diameter of the trap to ¼" still high losses of higher boiling alcohols were observed at -20 °C (Figures 3-4 and 3-5), and at -40 °C losses of the lower boiling alcohols were observed as well. Similar behaviour was found for 1/8" PFA, for which the recoveries for alcohols were even lower (Figures 3-6 and 3-7). With the exception of the alcohols, this method could be an option, but further testing of other methods for water removal was conducted to be able to perform better for this important group of compounds.

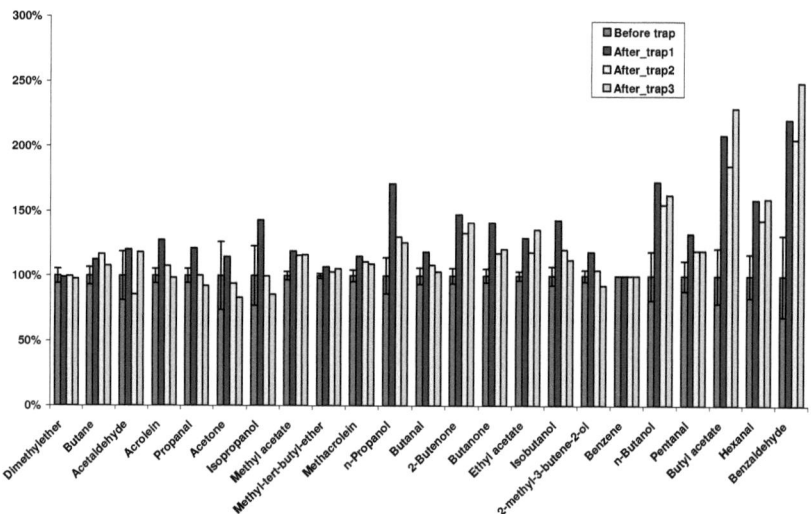

Figure 3-2. Measurements before and after silcosteel (1/8") cold trap at – 20°C. (For trap properties see Table 3-1).

3. Measurement of organic gases in ambient air and instrument development

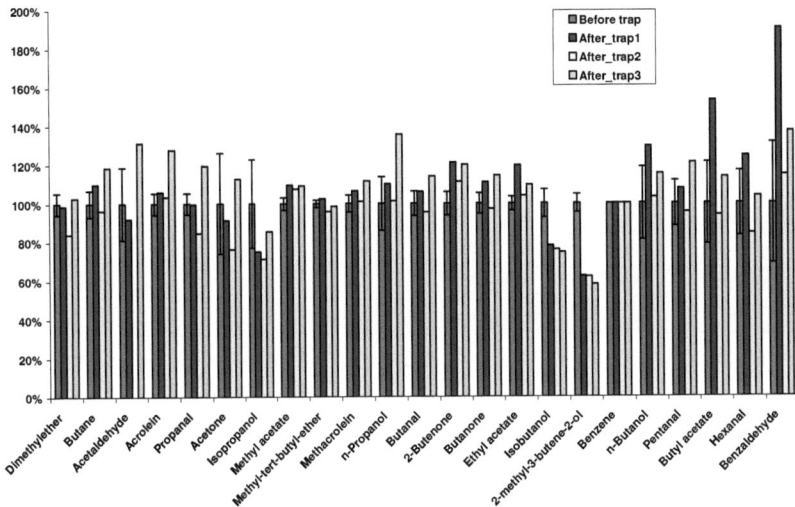

Figure 3-3. Measurements before and after silcosteel (1/8") cold trap at − 40°C. (For trap properties see Table 3-1).

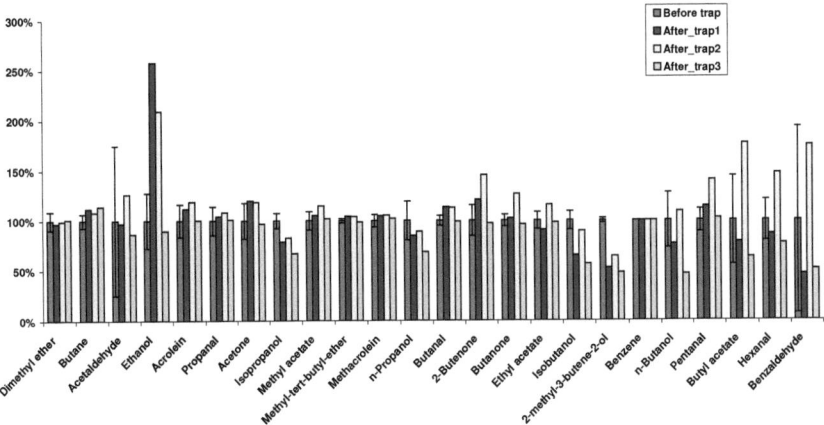

Figure 3-4. Measurements before and after silcosteel (1/4") cold trap at − 20°C. (For trap properties see Table 3-1).

3. Measurement of organic gases in ambient air and instrument development

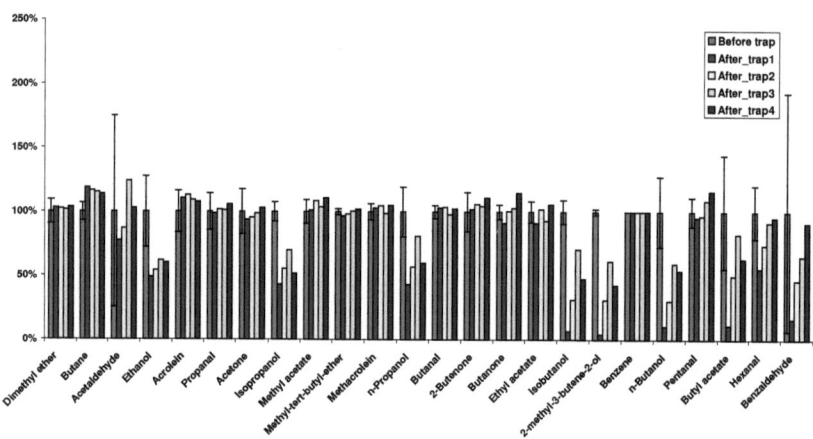

Figure 3-5. Measurements before and after silcosteel (1/4") cold trap at − 40°C. (For trap properties see Table 3-1).

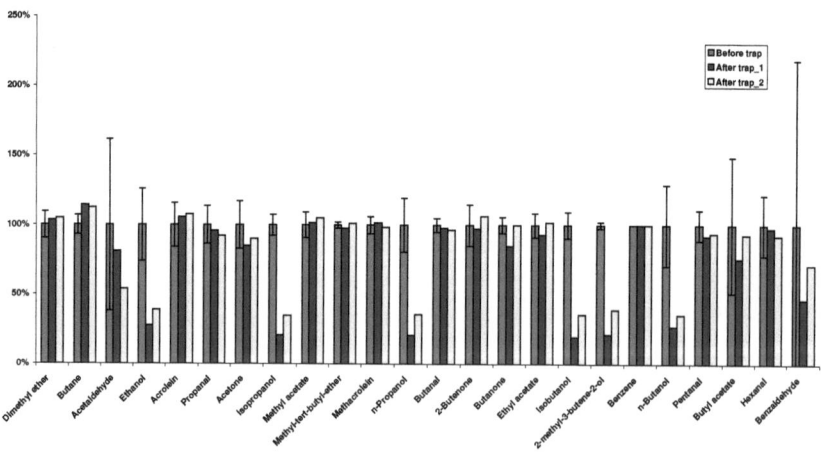

Figure 3-6. Measurements before and after PFA (1/8") cold trap at − 20°C. (For trap properties see Table 3-1).

3. Measurement of organic gases in ambient air and instrument development

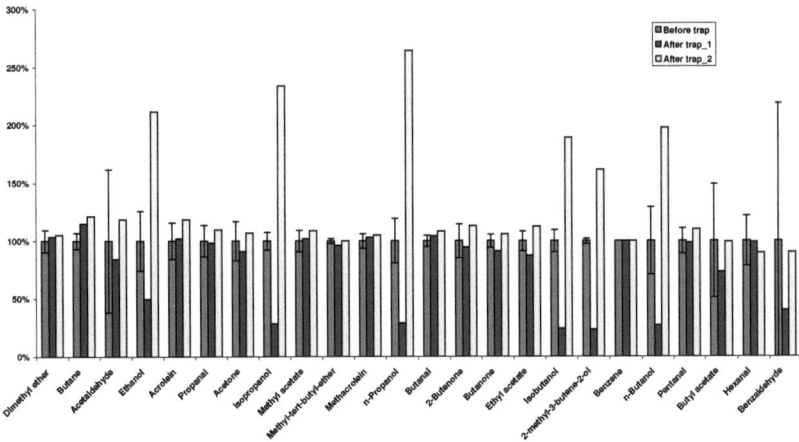

Figure 3-7. Measurements before and after PFA (1/8") cold trap at − 40°C. (For trap properties see Table 3-1).

3.3.2.2 Adsorptive removal of water

In this approach, a highly polar column (Carbowax Cp Wax 52CB) was used as a pre-column in order to remove the water before the trap. This is an efficient method, since by removing the water peak in the chromatogram the rest of the humidity in the sample is minimal. This method was described by Leibrock et al. (1997), who used a 10 m x 0.53 mm I.D. Carbowax column at 35 °C. We tested an identical column at 35 °C, 40 °C, 50 °C, 60 °C and 70 °C. The best separation between water and the analytes was obtained at 35 °C and 40 °C. Figure 3-8 shows the chromatograms of the OVOC standard mixed with humid air analysed isothermally at 35 °C. Since the peak from n-propanol and 2-methyl-3-buten-2-ol (MBO) partly overlapped with water it would be very difficult to prevent losses of these compounds. Especially if one considers the ageing of the column over time leading to peak shifts. Due to these problems it was chosen not to proceed with this method for water removal.

3. Measurement of organic gases in ambient air and instrument development

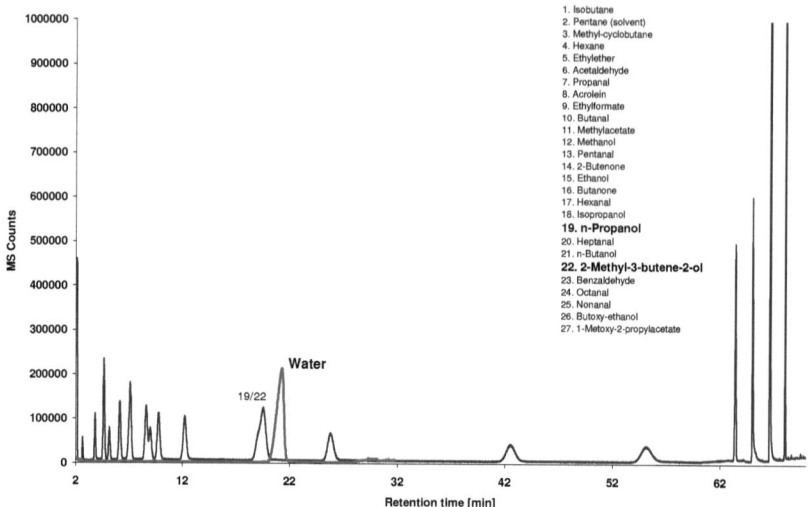

Figure 3-8. Chromatograms of humid OVOC standard for study of effect of adsorptive removal of water (see section 3.3.2.2)

3.3.2.3   Chromosorb W AW coated with lithium chloride

The trap was prepared by filling a 40 cm long 1/8" silcosteel tube with 0.41 g of LiCl coated Chromosorb W AW matrix. The matrix was held in place by silanized glass wool (Supelco). The 15% LiCl by weight gave an effective amount of 61.5 mg of LiCl, which, when sampling at room temperature, should remove the humidity from 1.5 L of air at 80% RH. An experiment to measure the rest humidity in the air sample after this trap was set up by coupling an air source in series with the trap and a humidity meter (Xentuar Portable Dewpoint Meter) (figure 3-9). The rest humidity after a Nafion tube was also measured as reference.

Figure 3-9. Set-up for testing the humidity of different air sources.

The various air sources were:
• Zero Air Can (Restek canister) – clean air produced from the in-house dilution system that utilises a Charcoal and Molecular Sieve trap cooled to - 80 °C to strip any impurities from an air cylinder (Air Products – zero plus grade).

3. Measurement of organic gases in ambient air and instrument development

- Lab Zero Air – clean air produced from a commercial generator (Parker Balston Zero Air Generator) with dryer unit fitted.
- Apel Zero Air Generator (ZAG) – clean air produced from passing ambient air over platinum catalyst at 400 °C with no dryer unit. Moisture content is essentially unchanged from ambient levels.

The dew point was measured both before and after the trap and the obtained results are shown in Table 3-2. To optimize the drying effect, a low flow rate was chosen for the test. The humidity sensor did not show any significant dependence on the flow rate. The zero air can from Restek was not tested further due to the low water content. The Nafion tube achieved the best drying effect, but as tests at Empa have shown, there are large losses of OVOCs on this water trap. The LiCl is not as efficient as the Nafion tube, but still removes the most of the humidity in the sample. Repeated testing of the LiCl trap showed the reusability of this trap, which after drying showed identical water removal efficiency.

Air source	Drying method	Flow rate	Dew point (°C)	Dew point
Zero air can (Restek)	-	2 l/min	-101.0	0.011 ppm
Lab zero air	-	2 l/min	-21.5	890 ppm
Lab zero air	Nafion	50 ml/min	-41.1	110.6 ppm
Apel ZAG	-	2 l/min	-6.8	3.46 ppb
Apel ZAG	-	1 l/min	-5.0	3.94 ppb
Apel ZAG	Nafion	50 ml/min	-37.0	176.8 ppm
Apel ZAG	LiCl	50 ml/min	-30.5	353 ppm

Table 3-2. Dew points of various air sources (with and without a drying process) measured with Xentuar portable dew point meter.

The next test involved the sampling of 1 litre of laboratory air on a cooled microtrap (-50 °C, HayeSep D, 15 mg), and analysis for water on the GC-MS. One sample was dried with Nafion dryer and two with the LiCl trap. The results are shown in figure 3-10, where the water peak for the air sample dried with LiCl trap is much larger than for the sample dried with Nafion dryer. Considering the results from the humidity meter this was not expected.

3. Measurement of organic gases in ambient air and instrument development

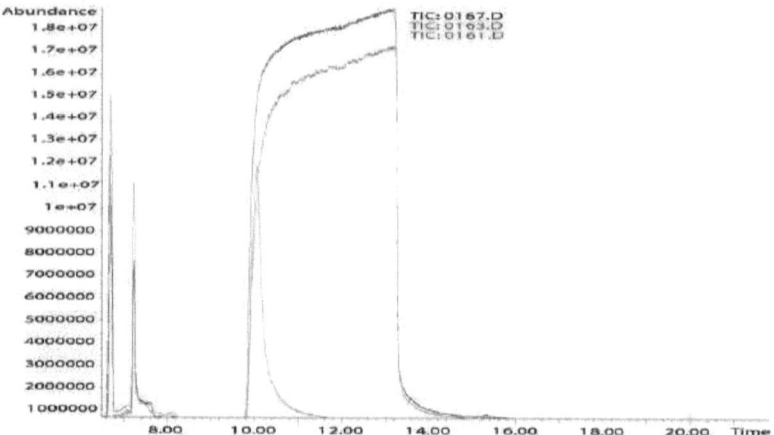

Figure 3-10. Chromatograms comparing water peak after drying air sample using LiCl (0167.D and 0161.D) and Nafion membrane (0163.D).

Due to the amount of water left after drying with the LiCl it was decided that this was not the method of choice, despite the initial promising results. Furthermore, losses of the compounds of interest at low concentration levels were reported as well (Monod et al. 2003).

3.3.2.4 Summary

All tested methods for the water removal had distinct problems. This resulted in further testing of water removal. One additional approach was to remove the water in the sampling step by using a hydrophobic adsorbent at room temperature. Tests at the University of Bristol showed promising results for Hayesep D, a polydivinybenzene (PDVB) polymere with a particle size 40-60 mesh. Based on these tests, this material was finally used. The bulk trap (see Figure 3.1) was kept at room temperature and was, after sampling, flushed with ~ 400 ml of dry helium for removal of the most part of the water. There was still a minimum of water left, which needed an analytical column with a certain degree of water resistance (see section 3.3.5).

3.3.3 Sampling artefacts by ozone

From the literature artefact formation from ozone in adsorbent sampling systems has been reported (Cao and Hewitt 1994; Helmig 1997). The ozone dependency of the MADS was tested by leading zero air through an ozone generator producing zero air with either 0 ppb, 48 ppb or 88 ppb ozone. With this zero air, the calibrations shown in figure 3-11 were performed for acetaldehyde, butanal and ethyl acetate on the MADS with Hayesep D as an adsorbent. There is a clear influence

3. Measurement of organic gases in ambient air and instrument development

from ozone on acetaldehyde, somewhat smaller influence on butanal and no influence on ethyl acetate. It was suspected that acetaldehyde and other oxygenates could be formed in the ozone generator as well, but this was not verified.

(a)
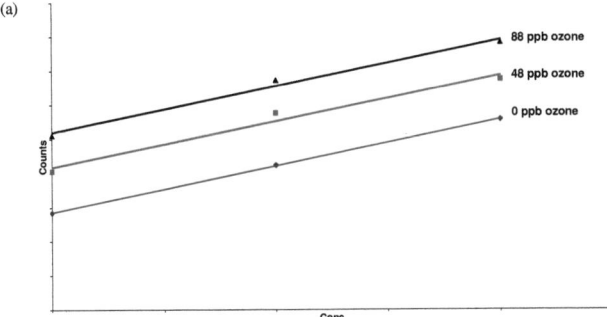

(b)
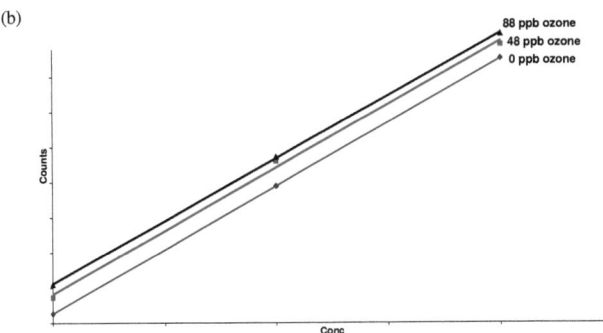

(c)
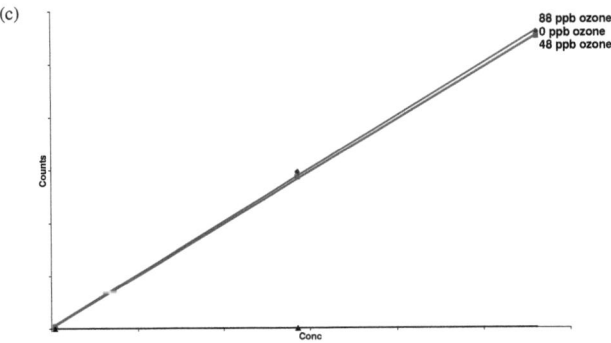

Figure 3-11. 3-Point calibration curve for acetaldehyde (a), butanal (b) and ethyl acetate (c) for three levels of ozone.

3. Measurement of organic gases in ambient air and instrument development

In a review Helmig (1997) described several techniques for the removal of ozone before sampling, concluding that the ozone removal technique must be chosen depending on the adsorbent used and the compounds of interest. During the instrument development we tested the following ozone scrubbers:

- Manganese dioxide filters.
- Potassium iodide (KI) coated traps, which were prepared by impregnating silanized glass wool (Restek) with KI in 15 cm ¼" PFA tubes.
- NO titration.

The first tests performed at the PTR-MS at PSI showed high losses of OVOCs on the MnO_2 filter, which excluded it from further testing. We continued with the KI filters and tested those under different conditions.

3.3.3.1 Potassium iodide ozone scrubber

The first tests of the KI ozone scrubber were performed at PSI on the PTR-MS. The aim was to observe if any losses due to the scrubber occurred. Figure 3-12 and 3-13 show that there were only small influences from the KI ozone scrubber on the recovery of the compounds. All values were calculated relative to benzene to compensate for drift in the mass spectrometer and pressure changes in the system due to the restriction from the scrubber. Recovery problems were only seen in the case of benzaldehyde at the GC-MS instrument, which was caused by losses of this compound on cold metal surfaces. This problem was discovered before starting the measurement campaigns and was solved by heating the reduction valve to 70 °C.

3. Measurement of organic gases in ambient air and instrument development

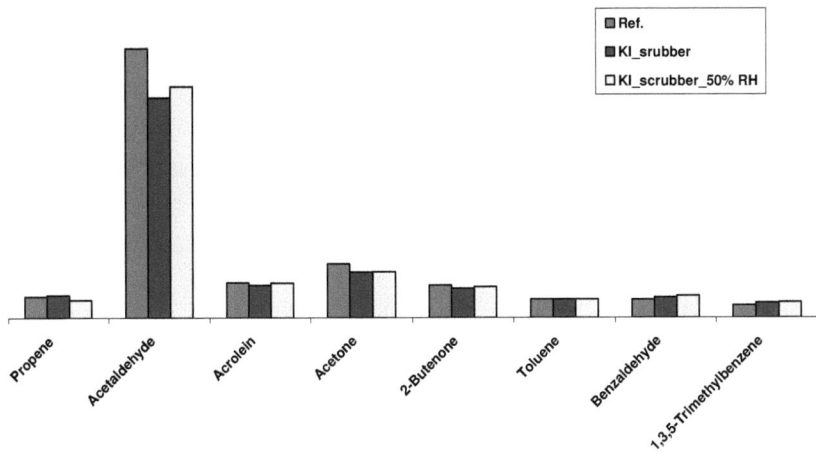

Figure 3-12. Test of losses over KI ozone scrubber performed on PTR-MS instrument at PSI. All values relative to toluene. For trap properties see text.

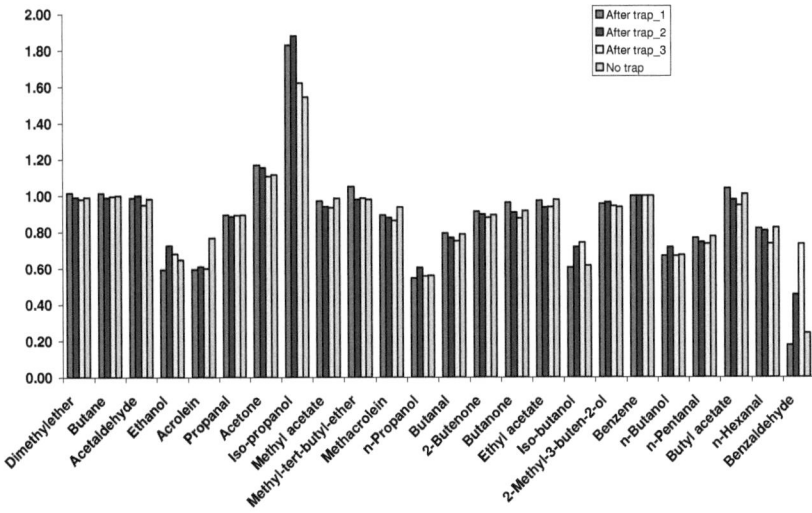

Figure 3-13. Test of KI ozone trap for losses performed on GC-MS system. Air mixture of 50% standard mixture and 50% humid zero air. All values relative to benzene. For trap properties see text.

3. Measurement of organic gases in ambient air and instrument development

These first results with low ozone concentrations were promising. Therefore, the KI-ozone scrubber was tested under real-air ozone conditions during the intercomparison at the SAPHIR smog chamber in Jülich. First blank experiments were performed without using the KI-scrubber. The smog chamber was filled with humid air containing 50 ppb of ozone. We spiked this air with standard and compared the calibration obtained with this zero air and with the humid zero air without ozone. Figure 3-14a and 3-14b show the results, which confirm the influence of ozone on the aldehydes but not on ethyl acetate.

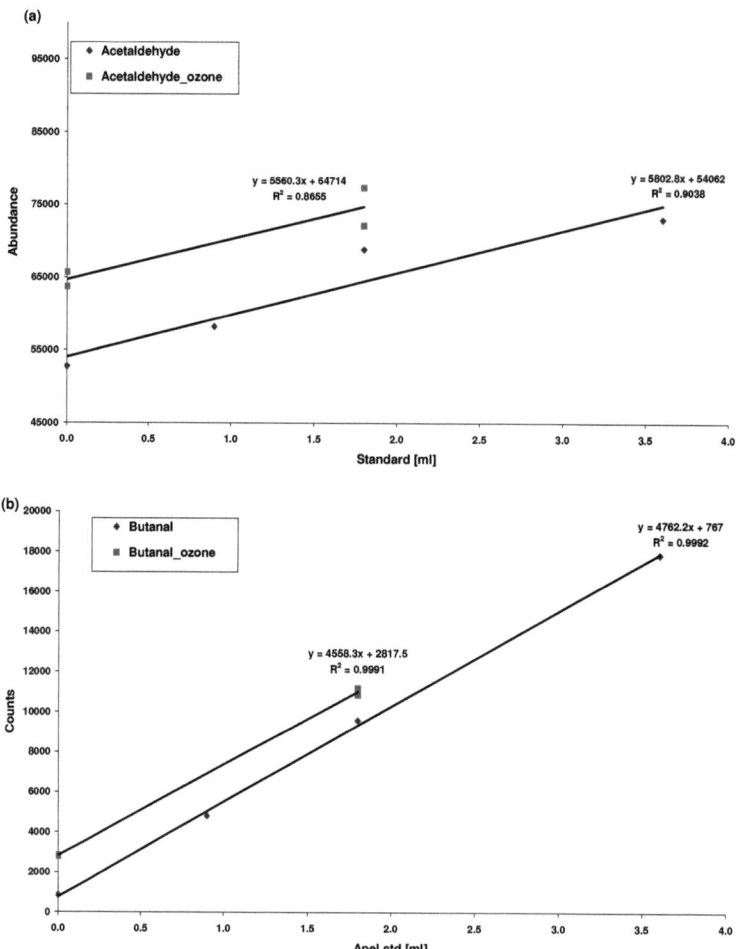

Figure 3-14a. 4-Point calibration curve for acetaldehyde (a) and butanal (b) for zero air overlaid with a 2-point calibration curve for zero air containing 50 ppb of ozone.

3. Measurement of organic gases in ambient air and instrument development

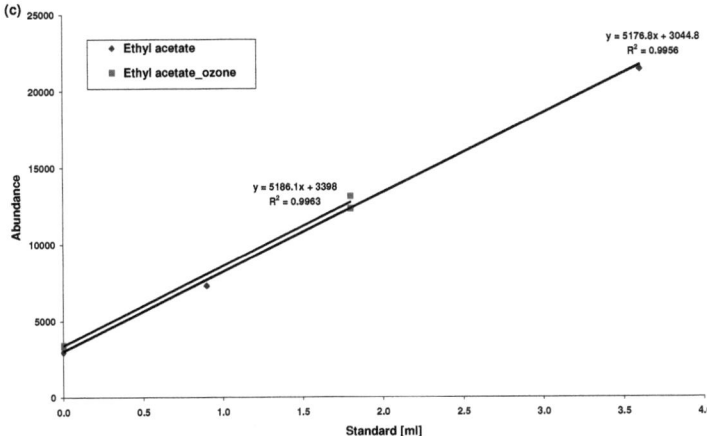

Figure 3-14b. 4-Point calibration curve for ethyl acetate (c) for zero air overlaid with a 2-point calibration curve for zero air containing 50 ppb of ozone.

In another part of the intercomparison experiment, humid zero air with 50 ppb ozone was spiked with 14 compounds. The last dilution step lasted long enough to analyse 3 parallels of the air with the KI-ozone scrubber, remove the scrubber, and analyse another 3 parallels. It should be noted that the mixing ratios continuously decreased since fresh zero air was injected into the chamber to compensate for the sampling losses. The results of this experiment are shown in Figure 3-15. For acetaldehyde, acetone and hexanal the mixing ratios increased after removing the ozone scrubber, which could imply that removal of ozone actually removed artefact formation for these compounds. This could also have been caused by adsorption on the ozone scrubber. Due to this uncertainty and also increased scatter in the data after introducing the KI ozone scrubber, it was chosen to proceed without this unit in the following campaigns and look carefully for artefact formation.

3. Measurement of organic gases in ambient air and instrument development

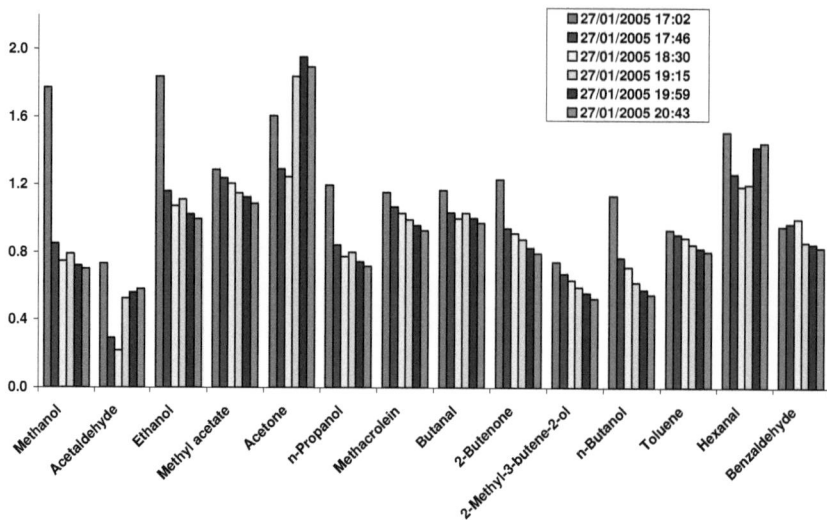

Figur 3-15. Test of influence of KI-ozone scrubber on results during low level concentrations with 50% RH and 50 ppb ozone in SAPHIRE chamber. Red line indicates the calculated reference concentration in the chamber.

3.3.3.2   Ozone removal by NO titration

The ozone removal by titration with nitrogen oxide (NO) is reported by Komenda et al. (2003). NO reacts fast with ozone and the produced NO_2 is considered to be non-reactive on the time scale it is present in the system. In the summer campaign of Zürich, we found that the aldehydes correlated extremely well with ozone. After personal communication with the R. Koppmann (ICG-II FZ Jülich, Germany), it was decided to test the NO titration for removal of ozone. Figure 3-16 shows the impact of ozone and NO titration on zero air (ZA), but this test was not straightforward to interpret due to suspected production of several oxygenated compounds in the ozone generator. The experiment was performed by flowing ZA through an ozone generator and connect it to the sample inlet of the MADS. After the ozone generator, the ¼" PFA sample line was connected with a T-piece to a mass flow controller, which was connected to a NO source. To ensure enough time for the reaction of ozone with NO, we installed a 5 meter sample line between the NO titration and MADS sample inlet. It was observed that NO titration had a positive effect on some compounds by removing ozone artefact production. This was also observed in the following measurement campaigns, where the correlations of several aldehydes with ozone were no longer observed. Therefore, after the 15th of July ozone was removed in the sampling step by NO titration.

3. Measurement of organic gases in ambient air and instrument development

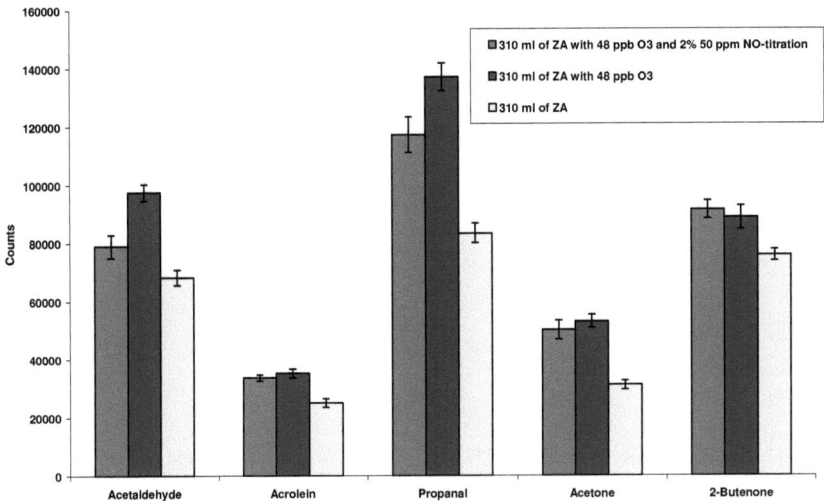

Figure 3.16. Effect of NO-titration on selected compounds. All values shown are relative to benzene. (ZA: zero air)

3.3.4 Adsorbent material

Adsorbent material tests were performed at the University of Bristol by D. Young. The main objectives were to test the adequacy of the adsorbents Hayesep D and HiSiv 3000 to sample OVOCs. Tests were performed on both adsorbent materials, but only the tests on Hayesep D will be reported since this was the material of choice for the MADS system. The reason for excluding HiSiv 3000 was its large difficulties with memory effects for some compounds (e.g. benzaldehyde). Hayesep D is a polydivinylbenzene (PDVB) polymer, and is mainly hydrophobic. The particle size used was 40-60 mesh. Figure 3-17 shows the breakthrough volumes of the material as a bulk trap with 2 g of adsorbent. In order to support the suitability of the method, the break through volumes of the most critical compounds, methanol and acetaldehyde, were tested. The bulk trap was able to collect almost 2 L of humidified air before methanol broke through at room temperature.

3. Measurement of organic gases in ambient air and instrument development

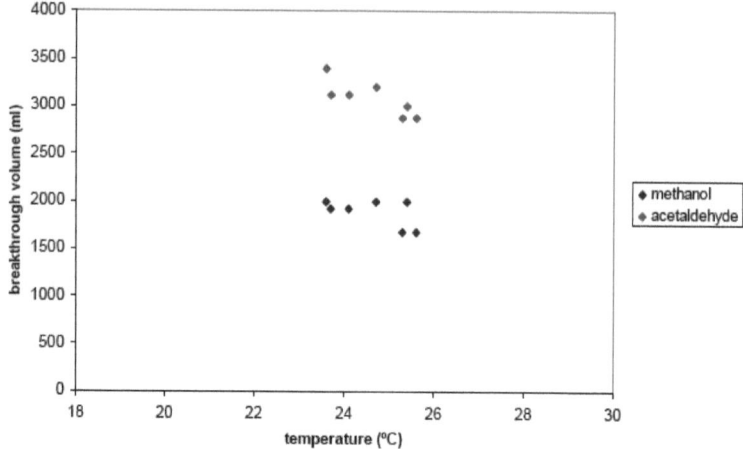

Figure 3-17. Breakthrough volumes on Hayesep D trap for methanol and acetaldehyde (2 g adsorbent, see text) at room temperature.

To ensure good separation of the compounds on the analytical column, a re-focusing step was required (see Figure 3-1). (The ADS system for halocarbons (Simmonds et al. 1995) used direct sampling onto a microtrap electrically cooled by peltier elements). This trap was tested and found suitable for this system as well. Figure 3-18 shows the breakthrough volumes of the microtrap at temperature down to –25 °C, and at the lowest temperature it was capable of collecting about 400 ml of OVOCs in dry helium before breakthrough occured.

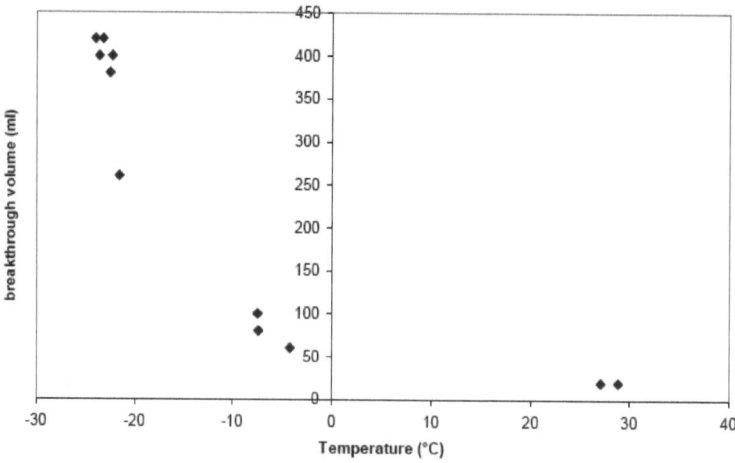

Figure 3-18. Breakthrough volumes on Hayesep D trap for methanol (14 mg adsorbent).

3. Measurement of organic gases in ambient air and instrument development

3.3.5 Separation

Several columns were tested for this instrument. The main challenge was to prevent tailing of polar compounds like the alcohols. An additional aspect was the water resistance and the possibility of good separation of the most important non-polar compounds as well. To combine these properties a semi-polar column would be the only choice. The main columns assessed were:

- ZB-624 (Zebron, 60 m, 0.32 mm, 1.8 μm film). This is a low bleed version of the popular "624" phase utilising 6% cyanopropyl-phenyl and 94% dimethylsiloxane.
- CP PoraBOND Q (Varian, 50 m, 0.32 mm, 5 μm film). This porous polymer PLOT column was the most stable column of its kind and withstood repeated water injections. The porous polymer is very pure and has virtually no catalytic activity. Retention times were repeatable, as the retention was not influenced by water in the sample. The reduction of bleed compared with a standard PLOT column was achieved by bonding the porous polymer layer and provides lower detection limits and faster stabilization with no random noise spikes.
- RT-Qplot (Restek, 30 m, 0.32 mm, 10 μm film). Bonded polystyrene-divinylbenzene based column specially developed for the separation of targeted apolar and polar compounds. Robust nature gives low bleed and minimal conditioning time. Polarity range between Porapak-Q and Porapak-N
- CP-PoraBOND U (Varian, 25 m x 0.32 mm, 7 μm film). This was a highly stable polar bonded porous polymer that eluted polar compounds as perfectly symmetrical peaks, allowing them to be analyzed together with light hydrocarbons or permanent gases. Retention times were repeatable, as the retention was not influenced by water in the sample. Low bleed was achieved in the same manner as for the PoraBOND Q column.
- CP-Wax 57 CB column (Varian, 10 m, 0,53 mm, 2 μm film). Chemically bonded polyethylene glycol with extensive cross-linking gave excellent temperature stability. High mid range polarity with good separation of alcohols, aldehydes and ketones.

Another aspect which came up during the testing was the ability of the column to refocus the compounds on the head of the column at 40°C, and the water resistance came to be very important for the lifetime of the column. The column which best fulfilled our requirements was the column CP Porabond U.

3.3.6 Detection by GC-MS and calibration

The detector was an Agilent HP 5973 mass selective detector with electron impact ionisation (EI), where the ionized molecules were separated and detected by a quadropole detector. The detector

3. Measurement of organic gases in ambient air and instrument development

was mostly running in selected ion monitoring (SIM) mode for maximum sensitivity. To ensure correct identification three characteristic masses were chosen for each molecule of interest. In the first air samples at each campaign, thorough investigation in scan mode was performed to identify possible interfering compounds with similar retention time.

The most commonly used calibration technique for OVOCs in air samples is dilution of a concentrated standard. There are two different techniques; either by static dilution in a chamber or by dynamic dilution into a stream of zero air (Boudries et al. 2002; Millet et al. 2004; Singh et al. 2004; Schade and Goldstein 2006). Another approach for the calibration is the use of diffusion tubes with concentrated standard solutions in an isolated chamber flushed with zero air at a set rate (Biesenthal et al. 1997; Leibrock and Slemr 1997; Hopkins et al. 2003). The pressure and temperature conditions must be carefully controlled for this technique to work properly. In our system the priorities were on the high mobility of the system, and the ability to add low amounts of standard. The first approach was to dilute a concentrated (~ 400 ppb) standard gas mixture (Apel & Riemer Environmental Inc., USA) into a 15 L silcocan canister (Restek). One concern of this method was the stability of the OVOCs in the canister. This was tested by adding Empa standard 1:100 diluted with pure nitrogen in the canister together with milli-Q water to obtain a 50 % RH. Figure 3-19 shows the measured mixing ratios after 27 days, 47 days and 84 days. The mixing ratios were clearly influenced by the contaminated "pure" nitrogen (6.0), which contained large amounts of acetaldehyde, acetone and other compounds. Since the mixture was clearly not stable enough in the canister, the dilution step had to be made on site. If the standard was not diluted the stability of the compounds in the Silcosteel canisters was excellent (Figure 3-20).

3. Measurement of organic gases in ambient air and instrument development

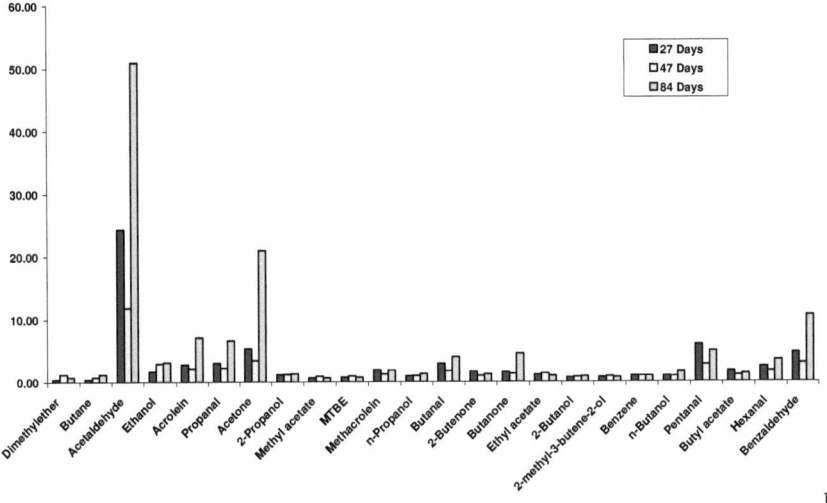

Figure 3-19. Stability of diluted standard mixture in silcocan canister. All values given as ratio to benzene.

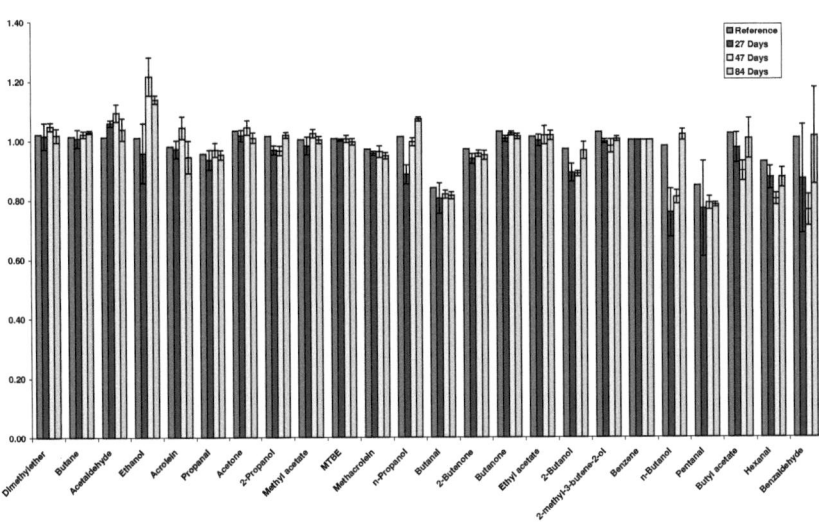

Figure 3-20. Stability of Empa standard mixture in silcocan canister. All values given as ratio to benzene.

Instead of transferring the standard mixture into smaller canister before each campaign, the decision was made to transport the Empa standard bottle (30 L) to each campaign site to prevent

3. Measurement of organic gases in ambient air and instrument development

additional sources of error. The standard was dynamically diluted by introducing an extra calibration valve in the system equipped with a 127 µl sample loop (Silcosteel (Restek)), that could be injected into either the sample stream or a stream of zero air or helium. This compensated for any losses of the samples after the calibration valve. The zero air generator was a Parker Zero air generator Model 1001 (Parker Inc.)

The standard gas was a multi-component mixture with 22 OVOCs plus butane and benzene. It was obtained from Apel & Riemer Environmental Inc., USA. A new standard was ordered and delivered at the end of September 2005 (see appendix II). Calibration curves obtained from this system are shown in Figure 3-21 during the ACCENT intercomparison. The problem of raised blank values for acetaldehyde will be discussed in section 3.4.2.

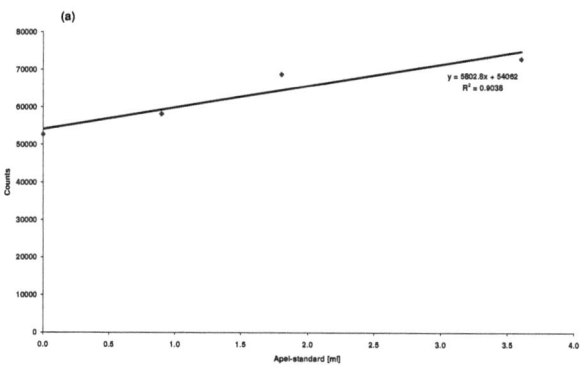

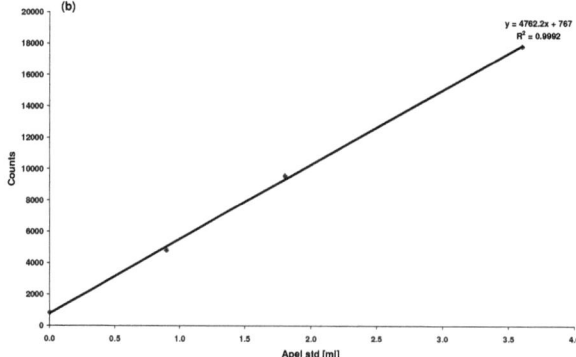

3. Measurement of organic gases in ambient air and instrument development

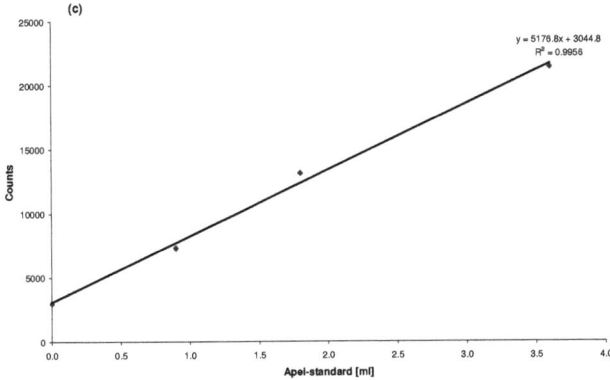

Figure 3-21. 4-Point calibration curve for acetaldehyde (a), butanal (b) and ethyl acetate (c) during the ACCENT intercomparison.

3.4 Summary of the developed OVOC instrument

The instrumental development in section 3.3 showed the difficulties in handling air samples for the analysis of OVOCs. Several problems were solved and in Figure 3-1 (see Section 3.3) the schematic overview of the final MADS coupled to the GC-MS is shown. The water removal was performed by collecting the samples on the relatively hydrophobic adsorbent bulk trap containing 0.6 g of 40 – 60 mesh Hayesep D at room temperature. After sampling, the adsorbent was flushed with 400 ml of dry helium to remove residual humidity. For removal of ozone 10 ml/min of 20 ppm NO was added to the air stream, as NO reacts rapidly with ozone, but not with the OVOCs (Komenda et al. 2003). To ensure good separation of the compounds, they were thermally desorbed (160 °C) onto a refocusing trap (30 mg Hayesep D) cooled to -40 °C with Peltier elements. The compounds were rapidly desorbed from this trap (180 °C) and transferred by a heated fused silica line into the gas chromatograph. The chromatographic separation was performed on a 25 m × 0.32 μm CP-Porabond U column (Varian Inc., USA) with 7 μm film thickness (carrier gas: helium, flow: 2 ml/min). Initially, the temperature was held at 40 °C for 2 minutes, and then increased rapidly to 120 °C at a rate of 20 °C/min, and then to 200 °C at a rate of 5 °C/min, where it was held constant until the end of the run. Individual compounds were detected by operating the mass spectrometer in single ion monitoring (SIM) mode, for an improved signal to noise ratio.

3. Measurement of organic gases in ambient air and instrument development

3.5 Intercomparison at the SAPHIR smog chamber

As a part of the European project ACCENT, subgroup Quality Assurance (QA), an intercomparison campaign was performed measuring OVOCs at the SAPHIR chamber at Forschungzentrum Jülich in January 2005. A total of 12 research groups were present to compare their latest methodologies for measuring the OVOCs. In Figure 3-21 the results from the last day of comparison are shown. On that day outside air was filled in the chamber and spiked with an unknown number of compounds. The results from the ENOVO instrument showed good agreement, although there seems to be a trend to underestimate the mixing ratios in the chamber. In general, the carbonyls and methyl acetate showed the most promising results (lower by 2,7% to 23,1%), and the alcohols seemed to be more problematic (lower by 19,3% to 43,4%). The losses for methanol were suspected to occur on the bulk trap in the water removal step, but no evidence for this was found. Further tests are required to solve this problem.

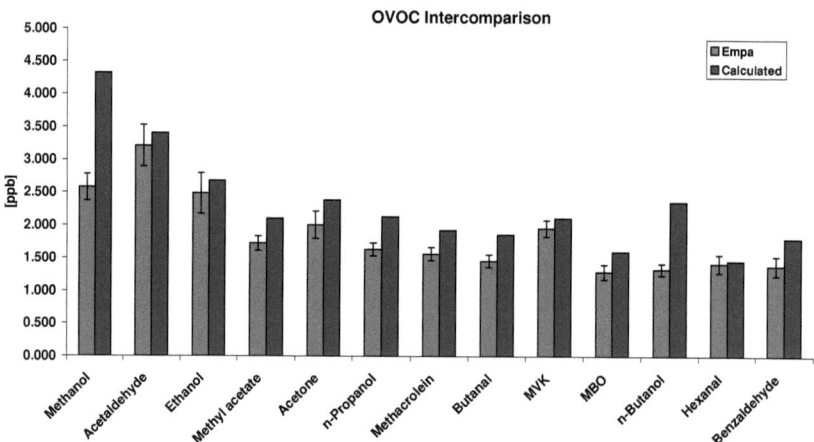

Figure 3-23. Results from intercomparison in SAPPHIRE smog chamber from Empa ADS GC-MS compared with concentrations added to the chamber.

3.6 Measurement campaigns

The first campaign was started in the middle of November 2004, where the instrument was mounted at the tunnel exit of a highway tunnel close to Zürich. After start-up problems due to the choice of the particle filter which adsorbed OVOCs, about 2,5 weeks of good quality data could be collected. Parallel measurements of CO, CO_2, traffic counts, and air velocity gave us the ability to estimate emission factors for vehicle classes and the whole vehicle fleet. The following campaign was an intercomparison of OVOC measurements at the SAPHIR smog-chamber in Jülich, Germany.

3. Measurement of organic gases in ambient air and instrument development

Following this campaign was the CLACE-4 campaign organized by Paul Scherrer Institute (PSI) in February 2005, which focused on ice cloud nuclei formation at the high alpine station Jungfraujoch. Thereafter, a total of four measurement campaigns were performed each at Jungfraujoch and at the urban background station in Kasernenhof, Zürich, the last campaign ended the first day in February 2006. In between, the instrument took part at a two week smog-chamber measurement at PSI in September 2005, and at the two week Aerowood campaign in a Swiss valley in December 2005 to measure residential wood burning emissions. Figure 3-23 gives an overview of the measurement during this PhD thesis.

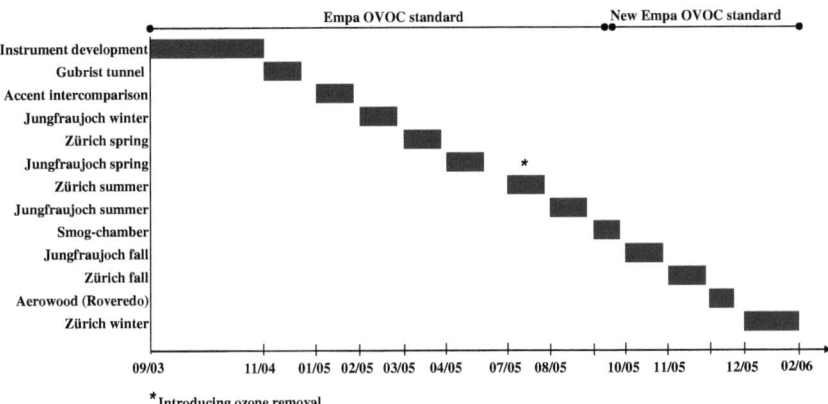

Figure 3-23. Overview over the two and a half year of method development and performed measurement campaigns.

3.7 Blank issues

The blank values for the compounds methanol, acetaldehyde, iso-propanol and acetone were high from the start of the instrument development. Iso-propanol was known to be applied in the cleaning of the Valcon E valve rotors, which we used for all valves in the MADS. This could be reduced by ultrasonic cleaning of rotors in water bath and Various tests were performed to identify the origin of the high background values for methanol, acetaldehyde and acetone. Other compounds were influenced as well. The following tests were done:

1) Blank runs with different desorption times from the bulk trap. The background values increased with the time the bulk trap is heated.

2) Blank runs with different temperatures for desorption. The higher the desorption temperature was, the more elevated was the background for the mentioned compounds.

3) Different amounts of ozone were added to the zero air. The background values increased with increasing amount of ozone.

3. Measurement of organic gases in ambient air and instrument development

4) Blank runs with different volumes of zero air (ZA). Also increasing amount of ZA increased the background. This means that oxygen or water also reacts with the Hayesep D adsorbent material.

5) From tests of ozone concentrations before at the inlet of the MADS and right before the bulk trap showed that ozone is not destroyed in the system. This implies that ozone could react on the adsorbent trap.

In the tests 1 to 4 it was observed that the background values increased with longer desorption time from the bulk trap, with higher desorption temperature, with increasing amounts of ozone and with increasing amounts of ZA. In addition, test 5 showed that ozone was not destroyed in the MADS before the bulk trap. From these results it was considered as quite certain that the raised background levels originated from the adsorbent material Hayesep D. One additional effect of the background was that it went down over time. In addition, when the system had been shut down for a while (more than a week), the background was elevated, and came down to previous values after a few days.

References:

Ammann, C., C. Spirig, et al. (2004). "Application of PTR-MS fro measurements of biogenic VOC in a deciduous forest." International Journal Of Mass Spectrometry **239**(2-3): 87-101.

Apel, E. C., A. J. Hills, et al. (2003). "A fast-GC/MS system to measure C-2 to C-4 carbonyls and methanol aboard aircraft." Journal of Geophysical Research-Atmospheres **108**(D20): art. no.-8794.

Biesenthal, T. A., Q. Wu, et al. (1997). "A study of relationships between isoprene, its oxidation products, and ozone, in the Lower Fraser Valley, BC." Atmospheric Environment **31**(14): 2049-2058.

Blake, R. S., K. P. Wyche, et al. (2006). "Chemical ionization reaction time-of-flight mass spectrometry: Multi-reagent analysis for determination of trace gas composition." International Journal Of Mass Spectrometry **254**(1-2): 85-93.

Boudries, H., J. W. Bottenheim, et al. (2002). "Distribution and trends of oxygenated hydrocarbons in the high Arctic derived from measurements in the atmospheric boundary layer and interstitial snow air during the ALERT2000 field campaign." Atmospheric Environment **36**(15-16): 2573-2583.

Cao, X. H., C.N. (1999). "The Sampling and Analysis of Volatile Organic Compounds in the Atmosphere." 119-157.

Cao, X. L. and C. N. Hewitt (1994). "Study Of The Degradation By Ozone Of Adsorbents And Of Hydrocarbons Adsorbed During The Passive Sampling Of Air." Environmental Science & Technology **28**(5): 757-762.

Dettmer, K. and W. Engewald (2003). "Ambient air analysis of volatile organic compounds using adsorptive enrichment." Chromatographia **57**: S339-S347.

Helmig, D. (1997). "Ozone removal techniques in the sampling of atmospheric volatile organic trace gases." Atmospheric Environment **31**(21): 3635-3651.

3. Measurement of organic gases in ambient air and instrument development

Helmig, D. (1999). "Air analysis by gas chromatography." <u>Journal Of Chromatography A</u> **843**(1-2): 129-146.

Helmig, D. and L. Vierling (1995). "Water-Adsorption Capacity Of The Solid Adsorbents Tenax-Ta, Tenax-Gr, Carbotrap, Carbotrap-C, Carbosieve-Siii, And Carboxen-569 And Water Management-Techniques For The Atmospheric Sampling Of Volatile Organic Trace Gases." <u>Analytical Chemistry</u> **67**(23): 4380-4386.

Hopkins, J. R., A. C. Lewis, et al. (2003). "A two-column method for long-term monitoring of non-methane hydrocarbons (NMHCs) and oxygenated volatile organic compounds (o-VOCs)." <u>Journal of Environmental Monitoring</u> **5**(1): 8-13.

Ioffe, B. V., V. A. Isidorov, et al. (1977). "Gas Chromatographic-Mass Spectrometric Determination Of Volatile Organic-Compounds In An Urban Atmosphere." <u>Journal Of Chromatography</u> **142**(NOV): 787-795.

Komenda, M., A. Schaub, et al. (2003). "Description and characterization of an on-line system for long- term measurements of isoprene, methyl vinyl ketone, and methacrolein in ambient air." <u>Journal of Chromatography A</u> **995**(1-2): 185-201.

Leibrock, E. and J. Slemr (1997). "Method for measurement of volatile oxygenated hydrocarbons in ambient air." <u>Atmospheric Environment</u> **31**(20): 3329-3339.

Millet, D. B., A. H. Goldstein, et al. (2004). "Volatile organic compound measurements at Trinidad Head, California, during ITCT 2K2: Analysis of sources, atmospheric composition, and aerosol residence times." <u>Journal Of Geophysical Research-Atmospheres</u> **109**(D23).

Monod, A., N. Bonnefoy, et al. (2003). "Methods for sampling and analysis of tropospheric ethanol in gaseous and aqueous phases." <u>Chemosphere</u> **52**(8): 1307-1319.

Riemer, D., W. Pos, et al. (1998). "Observations of nonmethane hydrocarbons and oxygenated volatile organic compounds at a rural site in the southeastern United States." <u>Journal Of Geophysical Research-Atmospheres</u> **103**(D21): 28111-28128.

Schade, G. W. and A. H. Goldstein (2006). "Seasonal measurements of acetone and methanol: Abundances and implications for atmospheric budgets." <u>Global Biogeochemical Cycles</u> **20**(1).

Simmonds, P. G., S. Odoherty, et al. (1995). "Automated Gas-Chromatograph Mass-Spectrometer For Routine Atmospheric Field-Measurements Of The Cfc Replacement Compounds, The Hydrofluorocarbons And Hydrochlorofluorocarbons." <u>Analytical Chemistry</u> **67**(4): 717-723.

Singh, H. B., L. J. Salas, et al. (2004). "Analysis of the atmospheric distribution, sources, and sinks of oxygenated volatile organic chemicals based on measurements over the Pacific during TRACE-P." <u>Journal of Geophysical Research-Atmospheres</u> **109**(D15): art. no.-D15S07.

Steinbacher, M. (2004). "Volatile organic compounds and their oxidation products in the atmospheric boundary layer: Laboratory and field experiments." Doctoral Thesis ETH No. **15557**.

Wisthaler, A., A. Hansel, et al. (2002). "Organic trace gas measurements by PTR-MS during INDOEX 1999." <u>Journal of Geophysical Research-Atmospheres</u> **107**(D19): art. no.-8024.

4 OVOC measurements in the Gubrist highway tunnel

Reprinted with permission from Legreid, G., S. Reimann, et al. (2007). "Measurements of OVOCs and NMHCs in a Swiss highway tunnel for estimation of road transport emissions." Environmental Science & Technology 41: 7060, Copyright 2012, American Chemical Society.

Measurements of OVOCs and NMHCs in a Swiss highway tunnel for estimation of road transport emissions

Geir Legreid[†,‡]*, Stefan Reimann*[*,†]*, Martin Steinbacher*[†]*, Johannes Staehelin*[§]*, Dickon Young*[II] *and Konrad Stemmler*[III]

[†] Empa, Swiss Federal Laboratories for Materials Testing and Research, Laboratory for Air Pollution and Environmental Technology,

Ueberlandstrasse 129, CH-8600 Duebendorf, Switzerland

[‡] Laboratory of Atmospheric Chemistry, Paul Scherrer Institut, CH-5232 Villigen PSI, Switzerland

[§] Institute for Atmospheric and Climate Science, Swiss Federal Institute of Technology, CH-8092 Zurich, Switzerland,

[II] School of Chemistry, University of Bristol, Cantock's Close, Bristol BS8 1TS, United Kingdom

[III] Laboratory of Radio and Environmental Chemistry, Paul Scherrer Institut, CH-5232 Villigen PSI, Switzerland

*Corresponding author phone: +41-44-8235511; fax: +41-44-621-6244;
e-mail: Stefan.reimann@empa.ch

4. OVOC measurements in the Gubrist highway tunnel

Abstract

Eighteen oxygenated volatile organic compounds (OVOCs) and eight non-methane hydrocarbons (NMHCs) were measured continuously during a two-week campaign in 2004 in the Gubrist highway tunnel (Switzerland). The study aimed to estimate selected OVOC and NMHC emissions of the current vehicle fleet under highway conditions.

For the measured OVOCs the highest EFs were found for ethanol (10 mg/km), isopropanol (3.2 mg/km), and acetaldehyde (2.5 mg/km), followed by acetone, benzaldehyde and acrolein. (Formaldehyde, the most abundant OVOC measured in other studies (1,2), was not measured by the method applied.) Relative emissions of the measured OVOCs were estimated to contribute approximately 6% and 4% to the total road traffic VOC emissions from Switzerland and Europe, respectively.

Results are compared with those from previous studies from the same tunnel performed in 1993 and 2002, and from campaigns in other tunnels. A continuous reduction in the emission factors (EFs) was determined for all measured compounds from 1993 until 2004. The relative contributions of light-duty vehicles (LDV) and heavy-duty vehicles (HDV) to the total emissions indicated that OVOCs were mainly produced by the HDVs, whereas LDVs dominated the production of the NMHCs.

1. Introduction

Road transportation is a major source of anthropogenic volatile organic compounds (VOCs) and nitrogen oxides (NO_x), which are key species in the processes leading to the production of tropospheric ozone and secondary organic aerosols (SOA) (3-5). In these processes oxygenated VOCs (OVOCs), which include carbonyls, ethers, alcohols, organic acids, organic peroxides and esters, play a crucial role both as primary and secondary pollutants. In addition, several OVOCs are toxic to human health and the environment, an example being acrolein, which is a highly toxic herbicide also produced during the combustion process (6).

Exhaust emissions from fossil-fuel driven vehicles are described by emission factors (EFs), which are defined as the emitted mass of a certain compound per driven distance. The EFs depend on many factors, such as fuel type, size and type of engine, driving conditions, and road gradient (7,8). EFs of single vehicles (or engines) can be determined from dynamometric or engine tests. This allows investigation of how driving cycles influence the emissions of different types of vehicles and provide the basis of road traffic emission models (9-14). Information on emissions from large and representative samples of vehicle fleets has been obtained from measurements in

4. OVOC measurements in the Gubrist highway tunnel

road tunnels (1,2,15-22) or at busy streets (23-25). These studies have been used to evaluate the results of road traffic emission models for particular conditions (mainly emissions for highway driving conditions) (26,27). Furthermore, road tunnel measurements can be used to investigate the impact of reduction measures and new technologies (i.e. catalytic converters) if repeated measurements are made in the same tunnel (2,19,22,28).

Contrary to legally confined compounds (i.e. NO_x, carbon monoxide (CO) and total VOCs), information on emissions of individual organic species is still incomplete. While reliable information on emissions of many NMHC species is available from dynamometric test measurements (9,29), data for OVOC species are limited. However, OVOC data are available from several tunnel campaigns, which were partially undertaken to study the impact of alternative fuels containing oxygenated additives such as ethanol and methyl-tert-butyl ether (MTBE) (2,15,17,30).

In this study, measurements of selected OVOCs and NMHCs, which were performed in 2004 in the Gubrist tunnel (Switzerland), are used to evaluate their emissions under highway conditions. Results are compared with those of other studies, in particular with OVOC data from the same tunnel in 1993 and NMHC data from 1993 and 2002.

2. Experimental

The Gubrist tunnel is part of the highway in the north-west of Zurich connecting three of the largest cities in Switzerland. It has one bore for each direction, with a total length of 3270 meters and a cross section of 48 m^2. There is a 100 km/h speed limit in the tunnel. The ventilation system was turned off during the measurements and the tunnel was passively ventilated through the piston effect of the passing vehicles. Two air velocity monitors were mounted 150 meters after the entrance and before the exit of the tunnel, providing 5 minute average values for the air velocity. Automatic loop detectors measured the traffic density, composition, and speed, averaged over 5 minute intervals. Sampling points for air samples were situated at both ends of the tunnel in the slightly ascending bore, with a road gradient of 1.3%. Samples for the OVOCs were taken only at the exit of the tunnel, 10 meters before the tunnel opening, and 1.5 meters above the road level. The entrance sampling point was situated 200 meters inside the tunnel, also 1.5 meters above the road level.

The main sampling line consisted of a 4 meter long ¼" PFA sampling tube equipped with a 1 µm PTFE supported inlet filter (Whatman Inc, USA) for removal of particulate matter, and was continuously flushed at 4 liters per minute. From the main sampling tube a 3 meter long ¼" PFA line was connected to the instrument. Every 50 minutes a sample of 45 ml was acquired over 4

4. OVOC measurements in the Gubrist highway tunnel

minutes. The samples were collected on a double adsorbent system connected to a gas chromatograph-mass spectrometer (GC-MS Agilent HP 6890-HP 5973N). The water removal was performed by collecting the samples on an adsorbent trap containing 0.6 g of 40 – 60 mesh Hayesep D, a polydivinylbenzene polymer, at room temperature. After sampling, the adsorbent was flushed with 400 ml of dry helium to remove residual humidity. To ensure good separation of the compounds, they were thermally desorbed (180 °C) onto a refocusing trap (30 mg Hayesep D) cooled to -40 °C with Peltier elements. The compounds were rapidly desorbed from this trap (180 °C) and transferred by a heated fused silica line onto the gas chromatograph. The chromatographic separation was performed on a 25 m × 0.32 µm CP-Porabond U column (Varian Inc., USA) with 7 µm film thickness. Initially, the temperature was held at 40 °C for 2 minutes, and then increased rapidly to 120 °C at a rate of 20 °C/min, and then to 200 °C at a rate of 5 °C/min, where it was held constant until the end of the run. Individual compounds were detected by operating the mass spectrometer in single ion monitoring (SIM) mode, for an improved signal to noise ratio. The compounds were identified by their mass spectra and quantified using a 24-compound OVOC standard gas mixture in the range of 350 – 450 ppb (Apel & Riemer Environmental Inc., USA), and a 30-compound VOC standard gas mixture in the range of 1 – 10 ppb (National Physical Laboratory, UK). Calibration was performed 3 times a day by filling a calibrated sample loop (127 µl), which was flushed with helium onto the adsorbent trap. The system was run over a time period of 20 days, with approximately 12 days of good quality data collected.

Further chemical analyses included CO by means of two commercial CO monitors based on the non-dispersive infrared (NDIR) technique (Horiba 350E). The CO monitors were mounted at both the exit and the entrance of the tunnel. The sampling frequency was 5 minutes. Carbon dioxide (CO_2) was monitored at the tunnel entrance and exit by two Li-6262 infrared H_2O/CO_2 analyzers (LI-COR). CO and CO_2 results will be published in a separate paper (Vollmer et al., in preparation).

The average emission factors were calculated from equation (1):

$$EF_{fleet} = \frac{\Delta C_p * V_{air} * t_{sample} * A_{tunnel}}{n_{vehicles} * l_{tunnel}} \quad (1)$$

where EF_{fleet} is the mean emission factor per vehicle for a given compound expressed as emitted mass per traveled distance, ΔC_p is the mixing ratio difference for a compound between the measurements at the tunnel exit and its entrance mixing ratio (see next paragraph), V_{air} is the air velocity in the tunnel, t_{sample} is the air sampling period for the chemical analyses, A_{tunnel} is the tunnel cross section, $n_{vehicles}$ is the number of vehicles passing the tunnel in the sampling period, and l_{tunnel} is the length of the tunnel between the sampling points. The data was synchronized by attributing each sample to the next 5 minute interval that gave the best overlap with the traffic data. For the calculation of the EFs only periods with traffic densities above 20 vehicles per 5 minutes and air velocity above 2 m/s. This was done to restrict the analysis to situations with a steady air flow and pollution transport in the tunnel. The EF was calculated for each sample, and all EFs above the mean plus three times the standard deviation were removed as outliers. The uncertainty was for each compound was defined as the 95% confidence interval for the EFs. Inherently, the uncertainty reported includes not only the analytical measurement error, but also the uncertainty caused by the statistical analysis including variation in fleet composition, driving pattern, and meteorological conditions.

Mixing ratios of the OVOCs at the entrance were estimated based on the assumption that the ratios between the OVOC and CO_2 mixing ratios at the entrance and at the exit of the tunnel were identical. This is reasonable since the mixing ratio at the entrance is largely influenced by road traffic, as the entrance sampling point is located about 200 meters inside the tunnel, and the air flow into the tunnel is mixed with air from the opposite tunnel bore. This assumption was supported by the three times larger mixing ratio of CO at the tunnel entrance compared to the CO values reported from a nearby urban location (Zürich). This allowed us to estimate the entrance mixing ratio for the VOCs ($C_{VOC,entrance}$) on a sample-to-sample basis using equation (2), where $C_{VOC,entrance}$ and $C_{VOC,exit}$ are the mixing ratios for the measured VOCs at both ends of the Gubrist tunnel, and $[CO_2]_{entrance}$ and $[CO_2]_{exit}$ are the CO_2 mixing ratios at the same locations:

$$C_{VOC,entrance} = \frac{C_{VOC,exit} * [CO_2]_{entrance}}{[CO_2]_{exit}} => \Delta Cp = C_{VOC,exit} - \left[\frac{C_{VOC,exit} * [CO_2]_{entrance}}{[CO_2]_{exit}}\right] \quad (2)$$

For the distinction between EFs for heavy duty vehicles (HDV) and light duty vehicles (LDV), a linear regression model according to Eq. (3) was used:

$$EF_{fleet}(t) = EF_{LDV}(1 - f_{HDV}) + EF_{HDV} f_{HDV}(t) + \varepsilon(t), \quad (3)$$

where f_{HDV} is the fraction of HDVs passing the tunnel in a given measurement interval and $\varepsilon(t)$ is the random error. The information on vehicle categories (LDV and HDV) was obtained from traffic count measurements with average values for every 5 minutes. LDVs are less than 6 meters long, which include small vans, cars and motorcycles, whereas HDVs are longer than 6 meters, including trucks, buses and larger vans.

3. Results and discussion

Figure 1 shows the air velocity, vehicle speed, traffic density and CO_2 mixing ratio at the tunnel exit. The air velocity was varying between 0 and 10.3 m/s and directly depended on the traffic density and the speed of the vehicles. During the commuter traffic hours the air velocity

4. OVOC measurements in the Gubrist highway tunnel

increased with the traffic density, and then slightly decreased when the traffic density was at its highest due to obstruction by the traffic. The maximum air velocity was reached during weekdays, due to the higher number of HDVs. At the weekend (4/12/04 and 5/12/04) the LDV counts were much lower, which is explained by the Swiss legislation banning goods transport during weekends. Only a small amount of buses were observed on these days. Since the proportion of diesel fuelled LDVs in Switzerland is low (2004: 9.1% (31)), most of the diesel emissions originated from the HDVs. The CO_2 mixing ratios at the tunnel exit also followed the traffic density with morning and afternoon peaks at weekdays. The maximum mixing ratio for CO_2 was lower on the weekend due to less traffic.

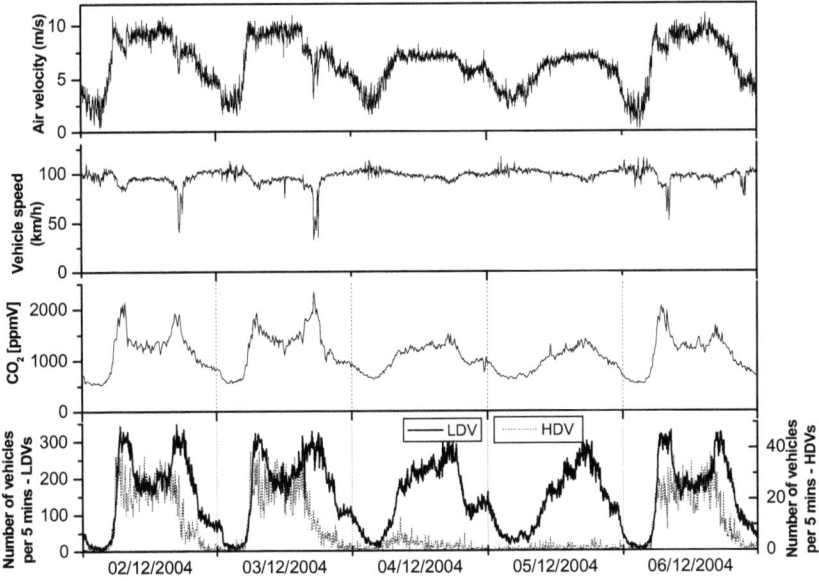

Figure 1. Air velocity, vehicle speed, CO_2 mixing ratio (Vollmer et al., in preparation), and traffic density at the tunnel exit. LDVs: Light Duty Vehicles (length < 6 m) and HDVs: Heavy Duty Vehicles (length > 6 m).

Figure 2 shows the measured mixing ratios of selected OVOCs (Panel A) and NMHCs (Panel B) at the exit of the Gubrist tunnel. Mixing ratios of OVOCs exhibited a diurnal cycle during the working days with a morning peak between 0630 and 0900 AM and an evening peak between 0400

4. OVOC measurements in the Gubrist highway tunnel

and 0630 PM. The lowest mixing ratios appeared as expected during the night. This clearly follows the pattern in the traffic counts shown in Figure 1, with low vehicle counts during night and increased vehicle density during commuter traffic hours. On weekend days the mixing ratios of the OVOCs showed a more gradual increase during the day and had a lower maximum value than during the week (Figure 2 – Panel A). For the NMHCs (Figure 2 – Panel B) the trend was similar, but with the difference that the maximum values on Saturday and Sunday were not significantly lower than during the week. This can be explained by the fact that the gasoline fuelled LDVs are mostly responsible for the NMHC emissions, in contrast to the OVOCs, which are mainly emitted by diesel fuelled HDVs (2,32) (also confirmed by this study – see below).

4. OVOC measurements in the Gubrist highway tunnel

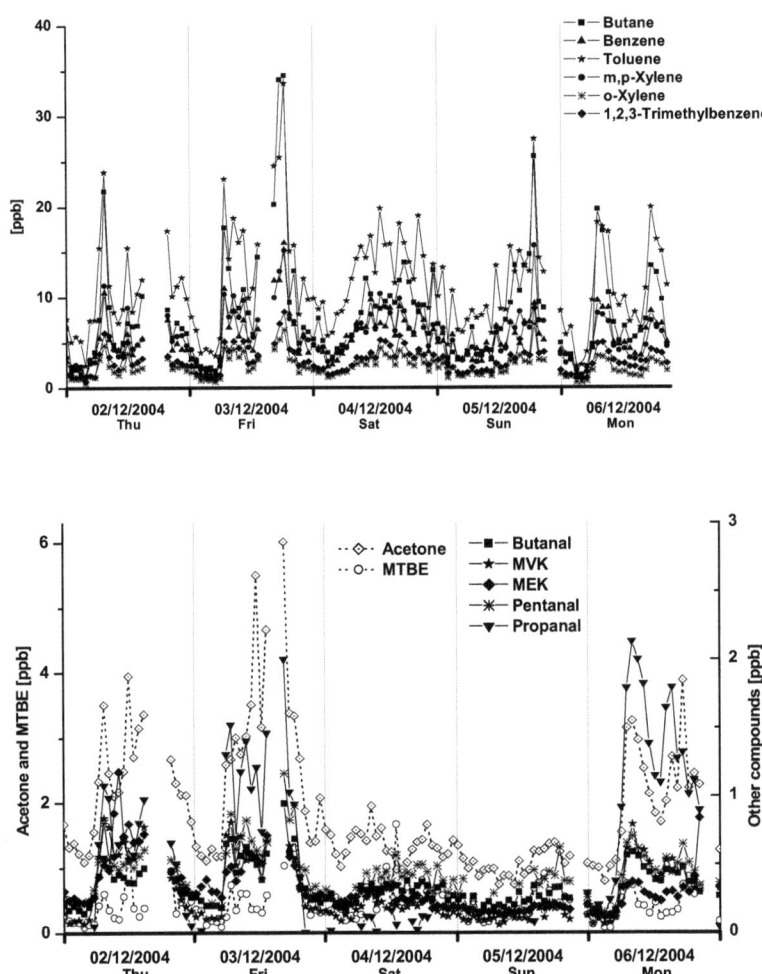

Figure 2. Mixing ratios for selected OVOCs (Panel A) and NMHCs (Panel B) at the exit of the Gubrist tunnel.

In table 1 the estimated entrance mixing ratios, calculated using the CO_2 method, are compared to measured mixing ratios from an urban background station in Zurich, which is within a distance of 10 km, in the same season. When comparing the two data sets using the interquartile range, most of the compounds showed estimated concentrations at the tunnel entrance to be higher in comparison with the measured concentrations in Zurich. This is reasonable since the entrance concentration is also influenced by the number of vehicles entering the tunnel and exiting the other tunnel bore. Exceptions are the esters, propanal, butanone (MEK) and acetone, whose potentially local industrial emissions in Zürich disturb the comparison. For propanal, MEK and acetone this could also be caused by the high tropospheric background mixing ratio, which makes it difficult to accurately determine the appropriate entrance mixing ratio. This would give an increased uncertainty for these compounds.

Compounds	Entrance mixing ratio estimated from CO_2 measurements (ppbV)		Urban background measured in Zürich one year later (ppbV)	
	1. Quartile	3. Quartile	1. Quartile	3. Quartile
Acetaldehyde	1.78	6.45	0.23	0.57
Propanal	0.08	0.40	0.08	0.13
Butanal	0.11	0.17	0.01	0.03
Pentanal	0.14	0.21	0.01	0.03
Hexanal	0.08	0.16	0.01	0.03
Benzaldehyde	0.42	0.60	0.01	0.02
Acrolein	0.41	0.87	0.04	0.11
MTBE	0.14	0.27	0.05	0.21
Acetone	0.70	1.21	0.71	1.57
MVK	0.09	0.21	0.02	0.04
MEK	0.10	0.19	0.10	0.23
Ethanol	4.57	15.81	2.61	9.22
Iso-propanol	1.03	3.64	0.45	1.49
n-Propanol	0.06	0.12	0.02	0.07
Iso-butanol	0.01	0.03	0.00	0.00
Methyl acetate	0.02	0.05	0.03	0.07
Ethyl acetate	0.04	0.14	0.11	0.35
Butyl acetate	0.06	0.12	0.06	0.19
Butane	2.04	4.16	0.73	1.97
1,3-Butadiene	0.65	1.09	0.03	0.13
Isoprene	0.40	0.80	0.03	0.11
Benzene	1.63	2.71	0.26	0.62
Toluene	2.92	5.72	0.74	2.42
m,p-Xylene	1.62	2.78	0.33	1.14
o-Xylene	0.66	1.15	0.13	0.42
1,2,4-Trimethylbenzene	0.99	1.55	0.02	0.07
	Measured mixing ratio at entrance (ppbV)		Parallel measurements in Zürich (ppbV)	
Carbon monoxide (CO)	0.962	1.873	0.303	0.682

Table 1. Mixing ratios of VOCs estimated from parallel CO_2 measurements at the entrance compared with the urban background in Zurich in the same season.

In Table 2 the average, maximum and minimum mixing ratios at the tunnel exit are shown. The compounds with the highest mixing ratios were ethanol and iso-propanol, followed by acetaldehyde and acetone. Formaldehyde, the most abundant OVOC measured in other studies (1,2), was not measured by the method applied. On wet, cold days high mixing ratios of ethanol and iso-propanol were found. Both of these compounds are major components of windscreen wiper fluids,

4. OVOC measurements in the Gubrist highway tunnel

and were therefore expected to be related to extensive windscreen cleaning. On these days mixing ratios of acetaldehyde and acetone were also up to 15 times higher than on dry days with similar traffic density. This might be caused by the enrichment in water droplets transported into the tunnel and evaporated during their residence time in the tunnel. Enrichment of acetaldehyde in rain and fog has been reported from several studies (e.g. Levsen et al. and Grosjean et al. (33,34)). This effect could also be responsible for the elevated acetone mixing ratios, as acetone has an even higher Henry's law constant (23.5 M/atm) than acetaldehyde (11.4 M/atm).

The most abundant of the quantified OVOCs, which is attributed exclusively to exhaust emissions, was acrolein, with an average mixing ratio at the tunnel exit of 1.39 ppb. In fact, acrolein is one of the most abundant aldehydes in the atmosphere (about 13% of the total atmospheric aldehydes (6)) and originates mostly from the combustion of fossil fuel. Of the total measured compounds in the Gubrist tunnel, the alcohols were the most prominent group of compounds comprising 54% of measurements, followed by the NMHCs with 28%, the aldehydes with 15%, and the ketones with 3%. The esters and MTBE represented each 0.4% of the measured compounds. The organic acids and difunctional carbonyls were not measured, but other studies found only small amounts of these compounds in exhaust gas (35,36).

Compounds	Concentration at tunnel exit [ppb]		
	Median	Mean	IQR
Acetaldehyde	9.1	14.3	4.3 - 12.8
Propanal	0.44	0.59	0.17 - 0.95
Butanal	0.28	0.30	0.21 - 0.37
Pentanal	0.38	0.40	0.28 - 0.49
Hexanal	0.25	0.27	0.17 - 0.37
Benzaldehyde	1.03	1.09	0.85 - 1.24
Acrolein	1.23	1.47	0.81 - 2.04
MTBE	0.37	0.47	0.26 - 0.58
Acetone	1.8	3.20	1.2 - 2.7
MVK	0.27	0.34	0.15 - 0.53
MEK	0.27	0.34	0.19 - 0.40
Ethanol	15.1	53.7	8.7 - 34.0
Iso-propanol	3.7	14.4	2.1 - 7.8
1-Propanol	0.15	0.29	0.11 - 0.25
Iso-butanol	0.03	0.07	0.02 - 0.07
Methyl acetate	0.06	0.08	0.03 - 0.09
Ethyl acetate	0.16	0.21	0.08 - 0.26
Butyl acetate	0.16	0.21	0.13 - 0.22
Butane	6.2	7.41	3.9 - 9.4
1,3-Butadiene	1.79	1.88	1.22 - 2.47
Isoprene	1.25	1.42	0.74 - 1.88
Benzene	4.5	4.81	3.1 - 6.1
Toluene	8.7	9.71	5.8 - 12.8
m,p-Xylene	4.4	4.86	3.1 - 6.1
o-Xylene	1.8	2.02	1.3 - 2.6
1,2,4-Trimethylbenzene	2.6	2.76	1.8 - 3.4

Table 2. Averaged, minimum and maximum mixing ratios at the exit of the Gubrist tunnel at working days. (bd: Below detection limit.)

4. OVOC measurements in the Gubrist highway tunnel

In table 3 the EF's for the measured compounds are listed together with EF's reported from other studies. The alcohols and the aromatic compounds dominated with 38% and 37% of the total EF's respectively. The aldehydes contributed 12%, the ketones 3.0% and the MTBE 0.6%. Ethanol had the highest single EF with 10 mg/km, followed by toluene (5.0 mg/km) and iso-propanol (3.3 mg/km). Of the other OVOCs, acetaldehyde was the most prominent with 2.4 mg/km, followed by acetone (0.78 mg/km) and benzaldehyde (0.64 mg/km).

Compared to the measurements performed in the same tunnel by Staehelin et al (1998) in 1993 (21), a large reduction could be found for the OVOCs and NMHCs. Acetaldehyde has been reduced by 30%, propanal by 63%, acrolein by 77%, and 2-butenone (MVK) by 25%. This is due to the improved engine technology for both LDVs and HDVs, and the increased fraction of LDVs with catalytic converters, up from approximately 67% in 1993 to 93% in 2002 and 95% in 2004 (31). The emission factors for the aromatic hydrocarbons (benzene, toluene, xylenes) in the Gubrist tunnel show a continued decreasing trend compared to those estimated from the same tunnel in 2002 (22) (-21% to -34%) and 1993 (-63% to -85%) (see figure 3). This is mainly due to the higher amount of Euro-3 and Euro-4 vehicles in the Swiss fleet in addition to the increased fraction of LDVs with catalytic converters, i.e. the continuous phase-out of large emitters with no catalytic converter. Butane shows a lower reduction (-14%) since 2002, which is caused by the fact that butane is the most volatile NMHC measured and therefore mostly emitted by evaporation and not through the exhaust of the vehicles. This is consistent with results from dynamometric measurements in a study based on European driving cycles by Ahlvik et al. (32), where about 4 times higher emissions of butane from evaporation than from exhaust gas were reported.

Compared to values reported from the tunnel study in Stockholm from 1999 (16), the emission factors for OVOCs and aromatic hydrocarbons were lower by 22% to 83% and 88% to 94%, respectively. The much smaller emissions in Switzerland are again expected to be mostly due to a different fleet composition, with more catalyst equipped vehicles in this study, 95%, compared to 60% in Stockholm. Reported values from a study in the Tauern tunnel in Austria in 1997 (60% catalyst-equipped cars) (19) are also higher than in the Gubrist tunnel in 2004, and the same trends are observed as for the Stockholm data with a higher difference in the emission factors for the aromatic compounds (-70% to -79%) than for the OVOCs (+6% to -68%). This can be related to the fact that diesel HDVs, which are mainly responsible for the OVOC emissions, have no catalytic converters.

The values reported from the USA are from two tunnels with totally different fleet composition, one with 0.1% HDVs (Caldecott) (1) and one with 60% HDVs (Tuscarora) (2). The emission factors from Caldecott are lower than the values reported from this study, due to the lower

4. OVOC measurements in the Gubrist highway tunnel

emissions of OVOCs from LDVs in comparison with HDVs. Benzaldehyde is the only exception with a larger emission factor in the Caldecott study. An explanation could be the lower toluene emissions from HDVs, as toluene is a precursor for benzaldehyde. The ratio of catalyst equipped LDVs from the study in Caldecott was 94%. A higher amount of HDVs might explain the elevated emission factors for the OVOCs in the Tuscarora tunnel. However, the EF for acetaldehyde and benzaldehyde were found to be higher in the Gubrist tunnel compared to the Tuscarora tunnel.

4. OVOC measurements in the Gubrist highway tunnel

Compounds	Gubrist 2004 (mg/km)	Number of measurements for EF-	Gubrist 2002 (mg/km)	Gubrist 1993 (mg/km)	Stockholm 1998/1999 (mg/km)	Caldecott, 1999 (mg/km)*	Tuscarora, 1999 (mg/km) (RSD)	Tauern 1997 (mg/km)**
Acetaldehyde	2.40 ± 0.39	138		3.44	3.1±0.8	1.31±0.14	2.19 (31)	2.7±0.1
Propanal	0.16 ± 0.02	156		0.45	0.47±0.2	0.14±0.02	0.33 (38)	
Butanal	0.12 ± 0.01	221			0.41±0.10	0.10±0.02	0.19 (50)	
Pentanal	0.18 ± 0.01	204				0.05±0.02	0.35 (37)	
Hexanal	0.13 ± 0.01	100			0.46±0.08	0.05±0.02	0.21 (74)	
Benzaldehyde	0.64 ± 0.04	202			3.8±0.7	0.81±0.10	0.44 (66)	0.6±n.a.
Acrolein	0.42 ± 0.05	97				0.19±0.02	0.41 (31)	0.6±0.0
MTBE	0.21 ± 0.01	214		1.82				
Acetone	0.78 ± 0.09	215		1.72	1.4±0.6	0.67±0.07	2.14 (38)	<2.1±n.a.
MVK	0.14 ± 0.01	219			0.2±0.5	0.10±0.02	0.46 (45)	
MEK	0.14 ± 0.01	220		0.18				
Ethanol	10.03 ± 2.07	214						
Iso-propanol	3.31 ± 0.74	201						
n-Propanol	0.09 ± 0.01	218						
Iso-butanol	0.03 ± 0.005	136						
Methyl acetate	0.03 ± 0.01	117						
Ethyl acetate	0.10 ± 0.01	205						
Butyl acetate	0.13 ± 0.01	193						
Butane	2.23 ± 0.10	219	2.6±0.4	11.28				
1,3-Butadiene	0.57 ± 0.02	182						
Isoprene	0.51 ± 0.03	181						
Benzene	2.13 ± 0.09	220	2.7±0.3	13.69	17.3±0.3			7.9±2.0
Toluene	4.99 ± 0.25	221	6.4±0.8	26.27	67.3±1.0			16.9±4.8
m,p-Xylene	2.91 ± 0.14	221	4.0±0.5	15.49	48.6±0.8			12.1±3.6
o-Xylene	1.20 ± 0.06	221	1.8±0.2	7.76	19.6±0.4			5.7±1.5
1,2,4-Trimethylbenzene	1.84 ± 0.07	221		5.04				
Road gradient	1.30%		1.30%	1.30%	-0.036	0.042	0-0.3%	1.50%
Average speed (km/h)	93		90	93	70	60	55	74
Average traction - heavy duty vehicles	5.5%		7%	9%	5%	0%	60%	18%

Table 3. Emission factors (EFs) from measurements in the Gubrist tunnel in 2004 compared with results from other studies. *Calculated from emission factors in mg/L assuming a LDV consumption of 14.2 L/ 100 km (39).

4. OVOC measurements in the Gubrist highway tunnel

The EFs separated for LDVs and HDVs by regression (see equation (3)) are shown in Table 4. Generally, the results of the EFs of HDVs have much larger uncertainties due to the low fraction of HDVs in the tunnel. The highest fraction of HDVs was 21%, and the extrapolation to 100% is connected with a substantial uncertainty. This explains the slightly negative (statistically insignificant) EFs for several compounds as also reported in other studies (19,21). For ethanol and iso-propanol the regression is not performed due to the high scatter in the data.

The results of the regression show that for many OVOCs the EFs of HDV are significantly larger than those of LDVs such as for propanal, butanal, pentanal, hexanal, acetone, MVK, MEK, ethyl acetate and butyl acetate. For some of the compounds this was also found by Staehelin et al. (21) from the measurements in 1993. Dynamometric tests have also shown higher emissions of OVOCs from HDVs in comparison to LDVs (35,36). MTBE is the only OVOC which has a significantly higher EF from the LDV, and is expected since this compound is added to gasoline (Swiss regulation: max. 15%, usual range: 2-10%). For some of the OVOCs the EFs are not significantly different due to high uncertainty for the HDV EFs. The highest uncertainty was found for acetaldehyde, which is caused by the previously discussed influence on cold, rainy days. For both vehicle classes (LDVs and HDVs) the OVOC emission factors in the Gubrist tunnel were lower in 2004 than in 1993. MEK is the only exception with approximately the same EF. In figure 3 the trend for LDV EFs for benzene and toluene is plotted with the reported values from the Gubrist tunnel (1993 and 2002) (22), from the Tauern tunnel in Austria (1988 and 1997) (19) and from this study. The main reduction was seen from 1988 to 1993, which is explained by the introduction of catalytic converters for the gasoline driven vehicles. There is even though still a reduction to be observed due to the already mentioned phase-out of older LDVs without catalytic converters and the increasing amount of Euro-3 and Euro-4 vehicles.

For all the measured NMHCs, except 1,3-butadiene, the regression results yield higher EFs from the LDVs than from the HDVs. This is also shown for the majority of compounds from dynamometric tests (35,36). However, the EFs for butane reported from these dynamometric tests were lower for the LDVs than for HDVs. The opposite result from the Gubrist tunnel is probably due to evaporation effects as mentioned earlier. The NMHC EFs of the LDVs and of the HDVs show the same decreasing trend from 1993 until 2004 as the EFs of the total fleet.

4. OVOC measurements in the Gubrist highway tunnel

Compounds	Gubrist 2004 (mg/km)		Gubrist 2002 (mg/km)		Gubrist 1993 (mg/km)	
	LDV	HDV	LDV	HDV	LDV	HDV
Acetaldehyde	2.1 ± 0.6	5.3 ± 6.6			2.3±1.1	14.6±5.1
Propanal	0.08 ± 0.03	1.10 ± 0.30			0.18±0.45	3.1±2.0
Butanal	0.11 ± 0.01	0.31 ± 0.11				
Pentanal	0.16 ± 0.01	0.45 ± 0.16				
Hexanal	0.07 ± 0.02	0.60 ± 0.19				
Benzaldehyde	0.64 ± 0.06	0.75 ± 0.70				
Acrolein	0.36 ± 0.07	0.71 ± 0.77			1.3±0.4	7.2±1.6
Methyl-t-butyl-ether	0.24 ± 0.02	-0.18 ± 0.24			0.06±0.05	0.27±0.25
Acetone	0.57 ± 0.13	3.94 ± 1.54			1.1±1.8	8.0±8.1
2-Butenone	0.08 ± 0.01	0.93 ± 0.13				
Butanone	0.11 ± 0.02	0.64 ± 0.19			0.08±0.44	1.2±2.0
Ethanol	6.6 ± 2.8	58.8 ± 33.7				
Iso-propanol	2.0 ± 0.4	2.5 ± 5.3				
1-Propanol	0.08 ± 0.02	0.32 ± 0.24				
Iso-butanol	0.02 ± 0.01	0.06 ± 0.08				
Methyl acetate	0.03 ± 0.01	0.08 ± 0.10				
Ethyl acetate	0.07 ± 0.02	0.52 ± 0.18				
Butyl acetate	0.12 ± 0.02	0.31 ± 0.20				
Butane	2.4 ± 0.2	-0.9 ± 1.9	2.7±0.3	0±3	9.7±5.3	27.3±27.1
1,3-Butadiene	0.56 ± 0.04	0.8 ± 0.5			1.6±0.2	-1.6±1.1
Isoprene	0.55 ± 0.04	0.0 ± 0.5				
Benzene	2.3 ± 0.1	-1.3 ± 1.8	3.3±0.2	0.7±1.6	10.3±6.2	20.9±34.1
Toluene	5.7 ± 0.4	-4.7 ± 5.0	8.7±0.5	1±5	20.4±6.9	33.1±35.0
m,p-Xylene	3.3 ± 0.2	-2.2 ± 2.6	4.2±0.3	1±3	10.8±3.0	27.2±15.3
o-Xylene	1.3 ± 0.1	-0.8 ± 1.1	1.9±0.1	0±2	4.8±0.6	6.3±2.9
1,2,4-Trimethylbenzene	2.0 ± 0.1	0.2 ± 1.4			4.6±1.1	9.6±5.5

Table 4. Emission factors for LDVs and HDVs determined by linear regression.

The Swiss and European (EU-15, i.e. the 15 countries belonging to EU before 2004) traffic emissions of the quantified OVOCs have been approximately estimated (Table 5) by using the ratio of the measured OVOCs with concurrently measured CO_2 at the tunnel exit. The slope for the linear regression line from the CO_2 and OVOC data was multiplied with the total CO_2 emissions from road transport for Switzerland (37) or EU-15 (38), respectively. Using this approach, it was assumed that the OVOC to CO_2 ratio represented the average fleet under highway driving conditions. This is most likely a lower limit for the emissions, since congested driving, as well as cold starts, are not included. All OVOC data were used for this estimation including data from days with elevated mixing ratios for some compounds due to the consumption of windscreen wiper fluid. For Switzerland the estimated OVOC emissions contribute to about 6% of the total road traffic VOC emissions, for EU-15 this fraction is about 4%. The main OVOCs are ethanol and isopropanol, the main components of windscreen wiper fluid, which account for more than 80% of the total OVOC emissions measured in this study. In 2004 the total LDV fleet in Switzerland was about 3.8 million vehicles, which gives an emission per vehicle of 0.6 liter of ethanol per year, assuming no other ethanol source.

4. OVOC measurements in the Gubrist highway tunnel

Compounds	Emission (kt/year)	
	Switzerland	EU-15
Acetaldehyde	0.10	5.54
Propanal	0.01	0.65
Butanal	0.01	0.37
Pentanal	0.01	0.58
Hexanal	0.01	0.30
Benzaldehyde	0.03	1.53
Acrolein	0.02	1.24
MTBE	0.02	0.89
Acetone	0.07	4.27
MVK	0.01	0.59
MEK	0.01	0.44
Ethanol	1.00	57.5
Isopropanol	0.37	21.4
Propanol	0.01	0.39
Ethyl acetate	0.01	0.33
Butyl acetate	0.01	0.39
Sum OVOC	1.68	96.5
Sum NMVOC	28.0	2555

Table 5. Estimate of the yearly road traffic emissions of the quantified OVOCs from measurements in the Gubrist tunnel, and the reported NMVOC values for the EU-15 countries and Switzerland.

The studies in the Gubrist tunnel have confirmed that the emissions of OVOCs and NMHCs from road transport have been reduced significantly with improved engine technology. There is still a need for more studies regarding the impact of the compounds in windscreen wiper fluids, and this might become even more important as emissions from combustion engines are expected to be reduced further.

Acknowledgement

We acknowledge the help of the staff at Amt für Abfall, Wasser, Energie und Luft (AWEL) for access to the tunnel infrastructure and for providing us with data for air velocity and traffic count. This study was financially supported by the Swiss Federal Office for the Environment (FOEN/BAFU). For helpful discussions we thank N. Heeb at Empa, P.G. Simmonds at the University of Bristol, and Robert A. Harley at the University of California, Berkely.

References:

(1) Kean, A. J.; Grosjean, E.; Grosjean, D.; Harley, R. A., On-road measurement of carbonyls in California light-duty vehicle emissions. *Environ. Sci. Technol.* **2001**, *35*, 4198-4204.

4. OVOC measurements in the Gubrist highway tunnel

(2) Grosjean, D.; Grosjean, E.; Gertler, A. W., On-road emissions of carbonyls from light-duty and heavy-duty vehicles. *Environ. Sci. Technol.* **2001**, *35*, 45-53.

(3) Millet, D. B.; Donahue, N. M.; Pandis, S. N.; Polidori, A.; Stanier, C. O.; Turpin, B. J.; Goldstein, A. H., Atmospheric volatile organic compound measurements during the Pittsburgh Air Quality Study: Results, interpretation, and quantification of primary and secondary contributions. *J. Geophys. Res.-Atmos.* **2005**, *110*.

(4) Odum, J. R.; Jungkamp, T. P. W.; Griffin, R. J.; Forstner, H. J. L.; Flagan, R. C.; Seinfeld, J. H., Aromatics, reformulated gasoline, and atmospheric organic aerosol formation. *Environ. Sci. Technol.* **1997**, *31*, 1890-1897.

(5) Odum, J. R.; Hoffmann, T.; Bowman, F.; Collins, D.; Flagan, R. C.; Seinfeld, J. H., Gas/particle partitioning and secondary organic aerosol yields. *Environ. Sci. Technol.* **1996**, *30*, 2580-2585.

(6) Ghilarducci, D. P.; Tjeerdema, R. S. Fate and effects of acrolein. In *Reviews Of Environmental Contamination And Toxicology, Vol 144*, 1995; Vol. 144, pp 95-146.

(7) BUWAL, Air Pollutant Emissions of Road Traffic 1950-2010. *Swiss Agency for the Environment, Forests and Landscape* **1995**.

(8) BUWAL, Air Pollutant Emissions of Road Traffic 1950-2010 (Supplement). *Swiss Agency for the Environment, Forests and Landscape* **1999**.

(9) Heeb, N. V.; Forss, A. M.; Saxer, C. J.; Wilhelm, P., Methane, benzene and alkyl benzene cold start emission data of gasoline-driven passenger cars representing the vehicle technology of the last two decades. *Atmos. Environ.* **2003**, *37*, 5185-5195.

(10) Weilenmann, M.; Soltic, P.; Saxer, C.; Forss, A. M.; Heeb, N., Regulated and nonregulated diesel and gasoline cold start emissions at different temperatures. *Atmos. Environ.* **2005**, *39*, 2433-2441.

(11) Mattrel, P.; Vasic, A. M.; Gujer, E.; Haag, R.; Weilenmann, M., VOC composition and ozone-forming potential of the exhaust gas of in-use motorcycles. *Int. J. Environm. Pollut.* **2004**, *22*, 301-311.

(12) Heeb, N. V.; Forss, A. M.; Weilenmann, M., Pre- and post-catalyst-, fuel-, velocity- and acceleration-dependent benzene emission data of gasoline-driven EURO-2 passenger cars and light duty vehicles. *Atmos. Environ.* **2002**, *36*, 4745-4756.

(13) Weilenmann, M.; Bach, C.; Rudy, C., Aspects of instantaneous emission measurement. *Int. J. Veh. Des.* **2001**, *27*, 94-104.

(14) Schmitz, T.; Hassel, D.; Weber, F. J., Determination of VOC-components in the exhaust of gasoline and diesel passenger cars. *Atmos. Environ.* **2000**, *34*, 4639-4647.

4. OVOC measurements in the Gubrist highway tunnel

(15) Kirchstetter, T. W.; Singer, B. C.; Harley, R. A.; Kendall, G. R.; Hesson, J. M., Impact of California reformulated gasoline on motor vehicle emissions. 2. Volatile organic compound speciation and reactivity. *Environ. Sci. Technol.* **1999**, *33*, 329-336.

(16) Kristensson, A.; Johansson, C.; Westerholm, R.; Swietlicki, E.; Gidhagen, L.; Wideqvist, U.; Vesely, V., Real-world traffic emission factors of gases and particles measured in a road tunnel in Stockholm, Sweden. *Atmos. Environ.* **2004**, *38*, 657-673.

(17) Lough, G. C.; Schauer, J. J.; Lonneman, W. A.; Allen, M. K., Summer and winter nonmethane hydrocarbon emissions from on-road motor vehicles in the Midwestern United States. *J. Air. Waste. Man. Assoc.* **2005**, *55*, 629-646.

(18) Sagebiel, J. C.; Zielinska, B.; Pierson, W. R.; Gertler, A. W., Real-world emissions and calculated reactivities of organic species from motor vehicles. *Atmos. Environ.* **1996**, *30*, 2287-2296.

(19) Schmid, H.; Pucher, E.; Ellinger, R.; Biebl, P.; Puxbaum, H., Decadal reductions of traffic emissions on a transit route in Austria - results of the Tauerntunnel experiment 1997. *Atmos. Environ.* **2001**, *35*, 3585-3593.

(20) Sjodin, A.; Persson, K.; Andreasson, K.; Arlander, B.; Galle, B., On-road emission factors derived from measurements in a traffic tunnel. *Int. J. Veh. Des.* **1998**, *20*, 147-158.

(21) Staehelin, J.; Keller, C.; Stahel, W.; Schlapfer, K.; Wunderli, S., Emission factors from road traffic from a tunnel study (Gubrist tunnel, Switzerland). Part III: Results of organic compounds, SO2 and speciation of organic exhaust emission. *Atmos. Environ.* **1998**, *32*, 999-1009.

(22) Stemmler, K.; Bugmann, S.; Buchmann, B.; Reimann, S.; Staehelin, J., Large decrease of VOC emissions of Switzerland's car fleet during the past decade: results from a highway tunnel study. *Atmos. Environ.* **2005**, *39*, 1009-1018.

(23) Ho, K. F.; Lee, S. C.; Chiu, G. M. Y., Characterization of selected volatile organic compounds, polycyclic aromatic hydrocarbons and carbonyl compounds at a roadside monitoring station. *Atmos. Environ.* **2002**, *36*, 57-65.

(24) Corsmeier, U.; Imhof, D.; Kohler, M.; Kuhlwein, J.; Kurtenbach, R.; Petrea, M.; Rosenbohm, E.; Vogel, B.; Vogt, U., Comparison of measured and model-calculated real-world traffic emissions. *Atmos. Environ.* **2005**, *39*, 5760-5775.

(25) Hueglin, C.; Buchmann, B.; Weber, R. O., Long-term observation of real-world road traffic emission factors on a motorway in Switzerland. *Atmos. Environ.* **2006**, *40*, 3696-3709.

(26) Hwa, M. Y.; Hsieh, C. C.; Wu, T. C.; Chang, L. F. W., Real-world vehicle emissions and VOCs profile in the Taipei tunnel located at Taiwan Taipei area. *Atmos. Environ.* **2002**, *36*, 1993-2002.

4. OVOC measurements in the Gubrist highway tunnel

(27) Sturm, P. J.; Rodler, J.; Lechner, B.; Almbauer, R. A., Validation of emission factors for road vehicles based on street tunnel measurements. *Int. J. Veh. Des.* **2001**, *27*, 65-75.

(28) Colberg, C. A.; Tona, B.; Stahel, W. A.; Meier, M.; Staehelin, J., Comparison of a road traffic emission model (HBEFA) with emissions derived from measurements in the Gubrist road tunnel, Switzerland. *Atmos. Environ.* **2005**, *39*, 4703-4714.

(29) Heeb, N. V.; Forss, A. M.; Bach, C.; Reimann, S.; Herzog, A.; Jackle, H. W., A comparison of benzene, toluene and C-2-benzenes mixing ratios in automotive exhaust and in the suburban atmosphere during the introduction of catalytic converter technology to the Swiss Car Fleet. *Atmos. Environ.* **2000**, *34*, 3103-3116.

(30) Kirchstetter, T. W.; Singer, B. C.; Harley, R. A.; Kendall, G. B.; Chan, W., Impact of oxygenated gasoline use on California light-duty vehicle emissions. *Environ. Sci. Technol.* **1996**, *30*, 661-670.

(31) BUWAL, Air Pollutant Emissions of Road Traffic 1980-2030. *Swiss Agency for the Environment, Forests and Landscape* **2004**.

(32) Ahlvik, P., Eggleston, S., Gorrisen, N., Hassel, D., Hickman, A.-J., Joumard, R., Ntziachristos, L., Rijkeboer, R., Samaras, Z., Zierock, K.-H., COPERT II "Computer program to calculate emissions from road transport". *Final draft report, April 1997, European Environment Agency* **1997**.

(33) Grosjean, D.; Wright, B., Carbonyls In Urban Fog, Ice Fog, Cloudwater And Rainwater. *Atmos. Environ.* **1983**, *17*, 2093-2096.

(34) Levsen, K.; Behnert, S.; Mussmann, P.; Raabe, M.; Priess, B., Organic-Compounds In-Cloud And Rain Water. *Int. J. Environm. Anal. Chem.* **1993**, *52*, 87-97.

(35) Schauer, J. J.; Kleeman, M. J.; Cass, G. R.; Simoneit, B. R. T., Measurement of emissions from air pollution sources. 5. C-1-C-32 organic compounds from gasoline-powered motor vehicles. *Environ. Sci. Technol.* **2002**, *36*, 1169-1180.

(36) Schauer, J. J.; Kleeman, M. J.; Cass, G. R.; Simoneit, B. R. T., Measurement of emissions from air pollution sources. 2. C-1 through C-30 organic compounds from medium duty diesel trucks. *Environ. Sci. Technol.* **1999**, *33*, 1578-1587.

(37) BUWAL, Climareporting. *http://umwelt-schweiz.ch/imperia/md/content/oekonomie/climareporting/inventar2003/inv03_05_03.pdf* **2005**.

(38) EMEP, Expert Emissions used in EMEP models. *http://webdab.emep.int* **2005**.

(39) Kean, A. J.; Harley, R. A.; Kendall, G. R., Effects of vehicle speed and engine load on motor vehicle emissions. *Environ. Sci. Technol.* **2003**, *37*, 3739-3746.

5 OVOC measurements in Zürich

Reprinted from Atmospheric Environment, 41, Legreid, G., J. Balzani Lööv, J. Staehelin, C. Hüglin, M. Hill, B. Buchmann, A. S. H. Prevot, and S. Reimann, "Oxygenated volatile organic compounds (OVOCs) at an urban background site in Zürich (Europe): Seasonal variation and source allocation", 8409–8423, Copyright (2012), with permission from Elsevier

Oxygenated Volatile Organic Compounds (OVOCs) at an urban background site in Zürich (Europe): Seasonal variation and source allocation

Geir Legreid[†,§], Jacob Balzani Lööv, Johannes Staehelin[‡], Christoph Hüglin[†], Matthias Hill[†], Brigitte Buchmann[†], Andre Prevot[‡] and Stefan Reimann[*,†,§]

[†] Empa, Swiss Federal Laboratories for Materials Testing and Research, Laboratory for Air Pollution and Environmental Technology,
Ueberlandstrasse 129, CH-8600 Duebendorf, Switzerland

[‡] Institute for Atmospheric and Climate Science, Swiss Federal Institute of Technology, CH-8092 Zürich, Switzerland,

[§] Laboratory of Atmospheric Chemistry, Paul Scherrer Institut, CH-5232 Villigen PSI, Switzerland

*Corresponding author phone: +41-44-823-5511; fax: +41-44-621-6244;
e-mail: stefan.reimann@empa.ch

Abstract

21 oxygenated volatile organic compounds (OVOCs) were measured in four seasonal campaigns at an urban background site in Zürich (Switzerland) with a newly developed double adsorbent sampling unit coupled to a gas chromatograph-mass spectrometer (GC-MS). In addition, selected non-methane hydrocarbons (NMHCs) were measured, as well as formaldehyde in the summer and winter campaign. The most abundant compound measured in all seasons was ethanol, with peak values of more than 60 ppb. Its seasonal variation with a lower mean value in summer compared to winter implied mostly anthropogenic sources. In contrast, compounds with additional biogenic sources, or compounds known to be produced in the troposphere by oxidation processes, had seasonal maxima in summer (e.g. methanol, acetone, formaldehyde, methacrolein and 2-butenone (methyl vinyl ketone, MVK)).

For the OVOCs it was estimated that local sources contributed 40 % and 49 % to the mixing ratios of the measured compounds in summer and in winter, respectively. Combustion was estimated to contribute with 75 % to these local sources independent of the season. About 50% of both the OVOC and NMHC levels in Zürich could be explained by the regional background, which included regional biogenic and anthropogenic sources in addition to secondary production. Industrial sources were identified for acetone, butanone (methyl ethyl ketone, MEK), n-propanol, iso-propanol, n-butanol, ethyl acetate and butyl acetate.

1. Introduction

Volatile organic compounds (VOCs) and nitrogen oxides (NOx) are important precursors for the production of ozone (O3) and secondary organic aerosols during summer smog episodes (Odum et al., 1996, 1997; Atkinson, 2000). Several primary as well as secondary air pollutants formed in summer smog are known to be toxic to humans, animals and plants, and efforts are required to reduce their concentrations (Altshuller, 1978; Heck et al., 1984). Oxidized VOCs (OVOCs) are an important fraction of the VOCs and are primarily emitted by vehicular transport, solvent usage, industry and biogenic sources (Placet et al., 2000; Sawyer et al., 2000; Legreid et al., 2007b). Furthermore, several OVOCs are produced by oxidation processes in the troposphere (Atkinson, 2000).

Carbonyls (aldehydes and ketones), being one of the most abundant group of OVOCs in urban air, have been measured in several studies during the last decades (Hewitt, 1999). Formaldehyde, acetaldehyde and acetone were reported to be among the most abundant

OVOCs in urban environments (Granby et al., 1997; Bakeas et al., 2003). Alcohols have not frequently been measured, but were included in urban campaigns in Boulder, USA (Goldan et al., 1995), in Stockholm, Sweden (Jonsson et al., 1985), in Sao Paolo, Brazil (Nguyen et al., 2001) and in Wuppertal, Germany (Niedojadlo, 2001). In Stockholm, large amounts of ethanol and iso-propanol were suspected to originate from window wiper fluid, whereas in Sao Paolo the influence of ethanol as replacement of gasoline in vehicles was studied. Niedojadlo (2001) also measured butyl acetate in Wuppertal, and discovered large industrial emissions. Furthermore, methyl-tert-butyl-ether (MTBE), a gasoline additive, has been detected in exhaust gas in studies from Europe and USA (Poulopoulos and Philippopoulos, 2000; Schade et al., 2002; Legreid et al., 2007b).

The most recent studies related to source identification of OVOCs and their emission strength in urban areas in Asia have been published by Zhao et al. (2004) and Kim et al. (2005), using factor analysis and positive matrix factorization, respectively. In Europe, Borbon et al. (2004) presented 4 years of measurement data of 50 non-methane hydrocarbons (NMHCs) and 17 OVOCs from a rural site in Eastern France, but the sample frequency was limited to two samples a week. For Zürich (Switzerland), Staehelin et al. (2001) used a principal component analysis (PCA) to determine the proportion of motor traffic emissions to the ambient air concentrations of 58 NMHCs. The data was collected from 1 year of continuous measurements, but no individual OVOCs were measured.

In this study a recently developed instrument for OVOC analysis (see Section 2.1) is applied. Section 3 contains the results from four one-monthly seasonal measurements of OVOCs and selected NMHCs from an urban background site in Zürich in 2005 and 2006. The measured compounds are discussed in relation to their relative abundances, as well as daily and annual cycles. In addition, anthropogenic sources are quantified by a source-tracer-ratio method (see Section 2.2) and their relative importance is discussed.

2. Methods

2.1 Measurements

The sample location in Zürich city centre is situated in a small park area (Kasernenhof) surrounded by busy streets within a residential area, and is regarded as an urban background site. Possible anthropogenic emissions from non-tailpipe OVOC sources include gasoline and solvent evaporation, residential heating and small industrial enterprises within the proximity

5. OVOC measurements in Zürich

of the site. At this site continuous measurements of O3, nitrogen oxides (NO+NO2), PM10 and sulphur dioxide (SO2) are performed by Empa within the Swiss national air pollution monitoring network (NABEL, 2006).

Air samples for OVOC analysis were taken from the main inlet of the measurement station by a 8 m x100 4 PFA sampling line equipped with a 0.5 mm silcotreated inlet filter (Restek Corp.) for removal of particulate matter. This line was continuously flushed with 0.5l of air per minute. In order to prevent producing aldehyde as an artefact during sampling (Northway et al., 2004), 10ml min—1 of 20 ppm NO was added to the air stream after 15 July 2005. Thereby, NO reacts rapidly with O3 but not with the OVOCs (Komenda et al., 2003) and the produced NO2 can be seen as non-reactive on the time scale of 0.5 min (i.e. the time which the air was in the sampling line). Every 50 min a sample of 350 ml was taken during 6 min.

The OVOCs were collected on a two-stage adsorbent system connected to a gas chromatograph with mass spectrometer (GC–MS Agilent HP 6890/HP 5973N). The water removal was performed on the sampling trap (0.6 g of Hayesep D (Supelco)) at room temperature. The hydrophobic nature of the adsorbent allowed most of the water to pass the trap, and the remaining humidity was removed by dry helium flushing. Thereafter, the compounds were refocused on a microtrap (14mg of Hayesep D) at —40 °C to improve the separation of the compounds on the analytical column. For further details, see Legreid et al. (2007a, b). The accuracy and precision calculated from intercomparison measurements were 3–25% (n-butanol: 37%) and 1–5%, respectively. This intercomparison was a part of the European project ACCENT, subgroup Quality Assurance (QA) and was performed at the SAPHIR smog chamber in January 2005. Methanol was only recovered at 60%, and this was corrected for in the following measurement campaigns. The detection limit for each compound was calculated as three times the standard deviation above the noise of five consecutive zero air samples. The sample error was calculated as the combined error of the given uncertainty in the calibration gas mixture and the uncertainty in the calibration. The latter was calculated from the analysis of five standards and five blanks and the reported value was calculated for the median value during the respective measurement campaign (Table 1). The calibration of the instrument was performed once per day by the analysis of a standard gas mixture (Apel & Riemer Environmental Inc., USA), with about 400 ppb mixing ratio of the compounds. A calibrated sample loop was used to dilute the standard with air from a zero-air generator (Parker Inc.). The system was in operation during 4 measurement campaigns in spring (18.03.05–18.04.05), summer (01.07.05–01.08.05), autumn (10.11.05–30.11.05) and winter (19.12.05– 01.02.06). After discarding data during instrument start-ups and during instrument failure, a total of almost 2500 samples were further analysed as described below.

In addition, formaldehyde has been measured using an instrument with the Hantzsch chemistry fluorimetric detection technique according to Junkermann and Burger (2006), a further development of the Aero Laser AL4021 (Hak et al., 2005). The calibration was performed by using liquid standards (range 3–4 ppb), and the gas phase concentration was calculated from the enrichment factor between gas and liquid flows in the stripper. For more details see Balzani (2007).

2.2 Source-tracer-ratio method

To estimate the contribution of the different sources to the abundance of OVOCs and NMHCs in Zürich, a source-tracer-ratio method was applied (Millet et al., 2005). This method assumes that there is a characteristic ratio for each compound to a source-specific tracer. To eliminate the influence of distant sources, a background subtraction was performed as follows. Firstly, the large-scale background was defined as the seasonal 0.1 quantile of the mixing ratio measured at the high alpine site Jungfraujoch. Measurements at Jungfraujoch were performed with the same instrument as in this study (Legreid et al., 2007a). Thereafter, the regional background was defined as the daily running 0.1 quantile and was subtracted from the data. The regional background represents the daily accumulation and is probably an overestimation. Data collected during two periods in winter with strong inversion were excluded from the source apportionment. The obtained data set was representative for the locally emitted pollutants, and the next step was to identify possible sources.

The sources taken into consideration to potentially contribute to atmospheric OVOCs were road traffic, residential heating, industry, evaporation and biogenic, as well as secondary production (i.e. oxidation of NMHCs). Road traffic and stationary combustion (i.e. room heating) could not be reasonably separated due to similar diurnal cycles and lack of specific tracers. This was also the case for secondary and biogenic sources due to similar influencing factors (temperature and radiation). The total concentration $[X]_i$ for each compound can then be explained as

$$[X]_i = [X]_{i,0} + [X]_{i,comb} + [X]_{i,industry} + [X]_{i,other} \tag{1}$$

where $[X]_{i,0}$ is the background mixing ratio (consisting of regional and large-scale background), $[X]_{i,comb}$ is the contribution of the combustion sources (mobile and stationary), $[X]_{i,industry}$ represents the industrial emissions and $[X]_{i,other}$ the other sources (biogenic, evaporative and other unidentified sources). Figure 1 provides an overview of the applied method.

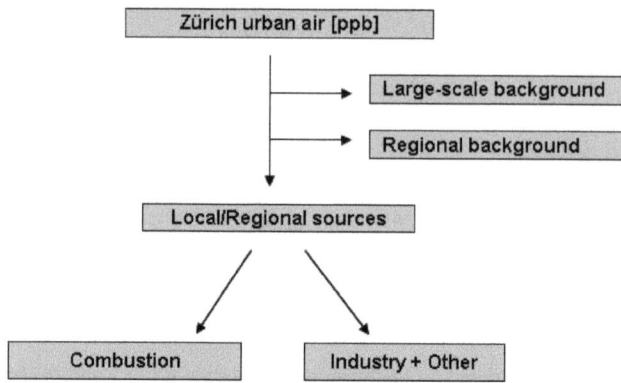

Figure 1. Step by step source allocation of summer data (compare text).

The contribution of combustion sources was determined using CO as the tracer. In Europe this compound is almost exclusively emitted from anthropogenic combustion processes and dominated by vehicle exhaust in urban regions in summer (Possanzini et al., 1996; Kuhlwein et al., 2002; EMEP, 2005). CO has a long tropospheric lifetime and is not significantly degraded within the timeframe of transport from the sources to the measurement site. All of the measured OVOCs are reported to be emitted from combustion sources (Schauer et al., 1999; McDonald et al., 2000; Schauer et al., 2001; Schauer et al., 2002). Assuming no other sources for CO, the contribution from combustion sources for each compound was calculated by an approach presented by Millet et al. (2005) using the equation (2). This was done separately for summer and winter data.

$$[X]_{i,comb} = [CO]_i \left(\frac{X}{CO}\right)_r \qquad (2)$$

where $(X/CO)_r$ is the characteristic emissions ratio for each compound X to CO after subtraction of background for both X and the tracer CO. However, this ratio cannot be

5. OVOC measurements in Zürich

retrieved by simple linear regression since the influence of other sources than combustion can bias the results. Instead, the ratio is obtained by varying the $(X/CO)_r$ over a range of possible values, and for each value the coefficient of determination (R^2) for the correlation of the residuals ($[X]_i - [X]_{i,0} - [X]_{i,comb}$) with CO is calculated. As an example, Figure 2 shows the R^2 for residual benzene plotted against the ratio $(Benzene/CO)_r$. The minimum in Figure 2 represents the optimum ratio $(Benzene/CO)_r$ describing the contribution from combustion sources.

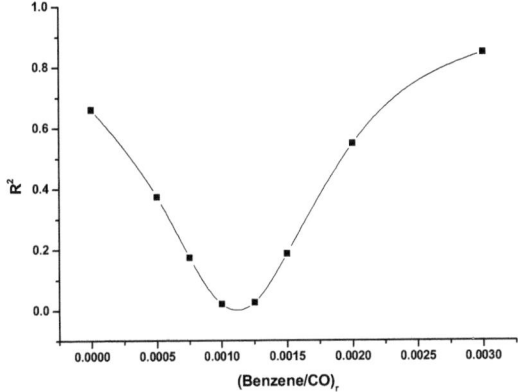

Figure 2. Plot for estimation of the combustion ratio to CO for benzene in winter applying the method described by Millet et al. (2005) (see text).

Compounds with a distinct difference in the ratio to CO on working days compared to weekends, were assumed to have substantial industrial sources (Steinbacher et al., 2005; Wang et al., 2005). In the case of Zürich the industries were assumed to be represented by nearby small-scale industries since there is no large emitting industry (such as chemical production or power plants) in the vicinity. As an example the weekday/weekend correlation between butyl acetate and CO is shown in Fig. 3. Similar behaviour was found for acetone, MEK, ethyl acetate, butyl acetate, iso-propanol, n-propanol and n-butanol. To obtain the characteristic emissions ratio (X/CO) for road traffic, only the weekend data were considered. The derived road traffic emissions ratio was then used to calculate the road traffic contribution on all days, and the excess concentration of the compound X was attributed to

5. OVOC measurements in Zürich

industrial sources. In summer enhanced mixing of the air masses and stronger evaporative, biogenic and secondary sources masked the weekday/weekend differences, and made it thereby impossible to isolate the industrial emissions. Therefore, during summer industrial sources have been included in the categories "combustion" and "other sources".

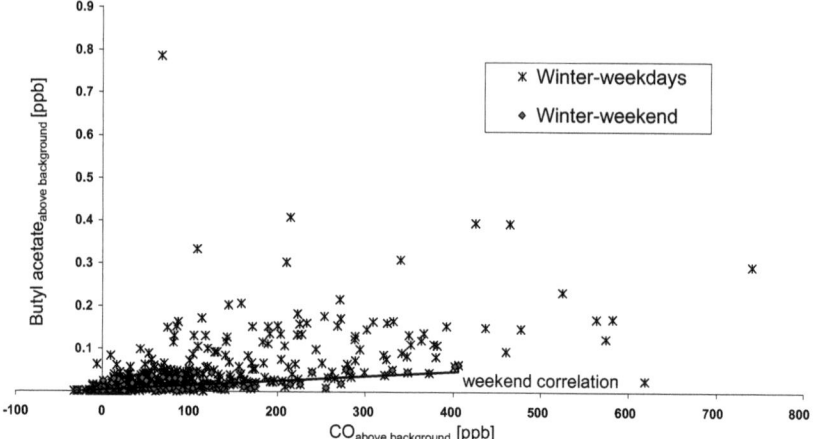

Figure 3. Butyl acetate mixing ratios versus carbon monoxide (CO) in winter. Data separated in weekdays and weekends. (Note that the daily running 0.1 quantile is subtracted for both butyl acetate and CO, which cause the negative values).

In summer the residual term was assumed to additionally originate from biogenic emissions, secondary production and evaporation, which all were incorporated in "other sources". Some of these sources would also to some extent be explained by the regional background. Note that the used procedure is not suitable to describe chemical degradation when important at local scale. This especially could impact compounds with very low tropospheric lifetimes.

3. Results and discussion

3.1 Seasonal variation

Table 1 the measured OVOCs and NMHCs from Zürich are listed as their seasonal median, mean and interquartile range (IQR: 25–75% quartiles) from the four seasonal campaigns. In total, 50% of the measured concentrations were alcohols, 16% aromatics, 15% aldehydes, 9% ketones, 2% esters, 7% butane/1,3-butadiene/isoprene and 1% MTBE. The most abundant OVOC measured was ethanol, which had the highest average annual mean mixing ratio of all the OVOCs with 6.6ppb and with maximum values of more than 60 ppb. Ethanol is a main component of window wiper fluids, and therefore our finding confirms results from studies performed in a Swiss tunnel (Legreid et al., 2007b) and in the city centre of Stockholm, Sweden (Jonsson et al., 1985). The second most abundant OVOC measured was formaldehyde (1.83 ppb), followed by methanol (1.80ppb) and acetone (1.45 ppb).

Methanol was the compound with the largest relative increase from winter to summer with 2.6 times higher median mixing ratio in summer (i.e. 1.21–3.18ppb). This can be explained by the strong biogenic sources for methanol (Jacob et al., 2005), which also accounts for the elevated mixing ratio for isoprene (2.5 x), hexanal (2.4 x) and acetone (2.0 x). Furthermore, products from the tropospheric oxidation of isoprene (Apel et al., 2002) also showed elevated mixing ratios in the summer compared with winter (i.e. methacrolein (2.0 x), methyl vinyl ketone (MVK) (2.0 x) and formaldehyde (1.2 x)). On the contrary, benzene, which is exclusively emitted from anthropogenic sources, showed much higher mixing ratios in winter compared to summer (3.4 times higher median mixing ratio). Larger winter values were also found for the mainly anthropogenic emitted compounds propanol (3.1 x), ethanol (2.3 x) and acrolein (2.3 x).

5. OVOC measurements in Zürich

	Spring 2005						Summer 2005						Fall 2005						Winter 2005/2006					
	Median	Mean	IQR	DTL	Error		Median	Mean	IQR	DTL	Error		Median	Mean	IQR	DTL	Error		Median	Mean	IQR	DTL	Error	
Formaldehyde	na	na	na - na	na	na		1.98	2.35	1.56 - 2.76	na	0.151		na	na	na - na	na	na		1.66	1.83	1.14 - 2.34	na	na	
Acetaldehyde	na	na	na - na	na	na		0.62	0.80	0.44 - 1.10	0.028	0.010		0.35	0.45	0.23 - 0.57	0.049	0.027		0.65	0.82	0.41 - 1.21	0.060	0.038	
Propanal	na	na	na - na	na	na		0.09	0.12	0.07 - 0.15	0.005	0.010		0.10	0.11	0.08 - 0.13	0.030	0.013		0.10	0.12	0.07 - 0.16	0.009	0.007	
Butanal	na	na	na - na	na	na		0.04	0.05	0.03 - 0.06	0.003	0.004		0.02	0.02	0.01 - 0.03	0.008	0.004		0.04	0.04	0.02 - 0.06	0.009	0.004	
Pentanal	na	na	na - na	na	na		0.03	0.03	0.02 - 0.04	0.002	0.003		0.02	0.02	0.01 - 0.03	0.003	0.002		0.02	0.02	0.01 - 0.03	0.009	0.003	
Hexanal	na	na	na - na	na	na		0.05	0.06	0.04 - 0.07	0.001	0.005		0.02	0.03	0.01 - 0.03	0.002	0.002		0.02	0.02	0.01 - 0.03	0.008	0.003	
Benzaldehyde	na	na	na - na	na	na		0.02	0.02	0.01 - 0.02	0.013	0.006		0.01	0.01	0.01 - 0.02	0.009	0.003		0.02	0.02	0.01 - 0.02	0.023	0.008	
Acrolein	0.07	0.08	0.05 - 0.10	0.009	0.002		0.05	0.06	0.03 - 0.07	0.004	0.006		0.06	0.09	0.04 - 0.11	0.009	0.006		0.12	0.14	0.07 - 0.21	0.007	0.008	
Methacrolein	0.01	0.01	0.01 - 0.02	0.002	0.000		0.03	0.04	0.02 - 0.05	0.001	0.003		0.01	0.01	0.01 - 0.01	0.001	0.001		0.01	0.02	0.01 - 0.02	0.003	0.001	
MTBE	0.10	0.14	0.06 - 0.17	0.000	0.001		0.13	0.18	0.09 - 0.22	0.000	0.011		0.09	0.15	0.05 - 0.21	0.000	0.007		0.07	0.09	0.04 - 0.11	0.000	0.004	
Acetone	1.45	1.66	0.96 - 2.12	0.060	0.013		1.81	2.12	1.40 - 2.55	0.034	0.137		0.95	1.24	0.71 - 1.57	0.040	0.058		0.88	1.17	0.60 - 1.62	0.017	0.046	
MVK	0.03	0.03	0.02 - 0.04	0.020	0.003		0.06	0.08	0.04 - 0.09	0.009	0.009		0.03	0.03	0.02 - 0.04	0.007	0.004		0.03	0.03	0.02 - 0.05	0.007	0.003	
MEK	0.19	0.24	0.14 - 0.32	0.002	0.002		0.16	0.20	0.11 - 0.25	0.001	0.011		0.15	0.17	0.10 - 0.23	0.004	0.011		0.16	0.22	0.10 - 0.32	0.003	0.009	
Methanol	1.83	2.18	1.26 - 2.84	0.121	0.043		3.05	3.18	2.15 - 3.97	0.061	0.391		0.86	1.11	0.56 - 1.33	0.086	0.066		0.97	1.21	0.67 - 1.62	0.102	0.045	
Ethanol	4.42	6.87	2.84 - 8.62	0.123	0.021		2.57	3.94	1.62 - 5.22	0.055	0.130		4.97	7.61	2.61 - 9.22	0.053	0.249		5.90	7.53	3.81 - 9.22	0.046	0.296	
Isopropanol	0.52	0.86	0.25 - 1.03	0.006	0.004		0.38	0.51	0.23 - 0.65	0.003	0.034		0.79	1.13	0.45 - 1.49	0.026	0.049		0.72	1.00	0.46 - 1.21	0.003	0.040	
Propanol	0.06	0.08	0.03 - 0.12	0.008	0.002		0.02	0.03	0.01 - 0.04	0.004	0.003		0.04	0.06	0.02 - 0.07	0.003	0.003		0.06	0.08	0.03 - 0.11	0.011	0.006	
MBO	0.06	0.08	0.04 - 0.10	0.000	0.001		0.08	0.11	0.05 - 0.14	0.000	0.006		0.13	0.18	0.08 - 0.23	0.007	0.009		0.14	0.16	0.08 - 0.21	0.000	0.005	
Butanol	0.01	0.01	0.01 - 0.01	0.020	0.003		0.02	0.07	0.01 - 0.04	0.011	0.004		0.03	0.04	0.02 - 0.05	0.000	0.002		0.02	0.06	0.01 - 0.03	0.013	0.021	
Methylacetate	0.05	0.06	0.04 - 0.07	0.004	0.001		0.04	0.05	0.03 - 0.06	0.003	0.005		0.05	0.06	0.03 - 0.07	0.000	0.003		0.06	0.07	0.04 - 0.11	0.006	0.004	
Ethylacetate	0.17	0.24	0.09 - 0.30	0.000	0.003		0.13	0.19	0.08 - 0.23	0.000	0.013		0.20	0.28	0.11 - 0.35	0.000	0.015		0.20	0.25	0.10 - 0.33	0.000	0.010	
Buthylacetate	0.07	0.10	0.03 - 0.13	0.000	0.001		0.04	0.06	0.02 - 0.08	0.000	0.004		0.10	0.15	0.06 - 0.19	0.000	0.006		0.08	0.13	0.04 - 0.18	0.003	0.005	
Butane	0.88	1.03	0.60 - 1.34	0.002	0.002		0.57	0.68	0.40 - 0.86	0.001	0.035		1.09	1.45	0.73 - 1.97	0.004	0.063		1.02	1.20	0.79 - 1.48	0.002	0.048	
1,3-Butadiene	0.06	0.07	0.04 - 0.09	0.000	0.001		0.06	0.07	0.04 - 0.09	0.001	0.002		0.06	0.08	0.03 - 0.13	0.000	0.004		0.08	0.09	0.05 - 0.12	0.000	0.003	
Isoprene	0.07	0.08	0.05 - 0.10	0.000	0.002		0.12	0.16	0.08 - 0.18	0.000	0.006		0.06	0.08	0.03 - 0.11	0.000	0.002		0.05	0.06	0.03 - 0.07	0.000	0.002	
Benzene	0.37	0.41	0.28 - 0.50	0.001	0.003		0.20	0.23	0.15 - 0.29	0.000	0.014		0.38	0.48	0.26 - 0.62	0.001	0.031		0.68	0.75	0.41 - 1.02	0.001	0.029	
Toluene	1.18	1.46	0.78 - 1.76	0.003	0.340		1.12	1.43	0.84 - 1.89	0.001	0.043		1.28	1.70	0.84 - 2.42	0.001	0.078		0.99	1.25	0.66 - 1.61	0.002	0.048	
Ethylbenzene	0.22	0.26	0.17 - 0.30	0.005	0.033		0.18	0.21	0.14 - 0.26	0.001	0.006		0.19	0.25	0.11 - 0.35	0.001	0.008		0.17	0.21	0.12 - 0.28	0.000	0.007	
m,p-Xylene	0.59	0.72	0.43 - 0.86	0.008	0.446		0.49	0.58	0.36 - 0.74	0.001	0.026		0.61	0.80	0.33 - 1.14	0.030	0.080		0.50	0.66	0.33 - 0.85	0.000	0.023	
o-Xylene	0.21	0.25	0.15 - 0.30	0.026	0.020		0.20	0.25	0.15 - 0.32	0.005	0.009		0.23	0.30	0.13 - 0.42	0.000	0.000		0.21	0.25	0.14 - 0.33	0.001	0.008	

Table 1. Mixing ratios measured in Zürich during seasonal campaigns in 2005/2006. The median, mean, interquartile range (IQR) and measurement error are listed for each season as also for all measurements. (IQR: Interquartile range, DTL: Detection limit, na: not available).

3.2 Diurnal variation

Fig. 4 shows the diurnal cycles of trace gases in Zürich, measured simultaneously with the OVOCs during the campaigns. For primarily emitted pollutants such as CO, SO2, benzene and toluene the anthropogenic emissions and meteorology determine the diurnal cycles. At night (until about 05:00), mixing with cleaner air masses from the surrounding areas and lower emissions reduce the mixing ratios. Chemical reactions and deposition can for some compounds (such as O3) play an important role as well. Thereafter, traffic emissions lead to local maxima in the morning, which is found for the anthropogenic pollutants NOx, benzene and toluene (mainly from traffic). The following rapid decrease is due to the break-up of the nighttime inversion and vertical mixing with the residual layer, which is enriched in O3 (see O3 diurnal cycle). In the evening the build-up of the inversion and continuous emissions into the urban air, lead to another maximum for the pollutants.

The morning peak in winter appears 1–2 h later than in summer due to later break-up of the inversion, which can also be identified in the O3 diurnal cycle. The 6 times higher morning peak increase for SO2 in winter compared to that in summer is an indication of stationary combustion sources. In Zürich oil burners are frequently used for room heating, and they are known to emit more SO2 than traffic sources due to the higher sulphur content in oil compared to gasoline. On the other hand, SO2 emissions in summer are probably mostly caused by road traffic.

5. OVOC measurements in Zürich

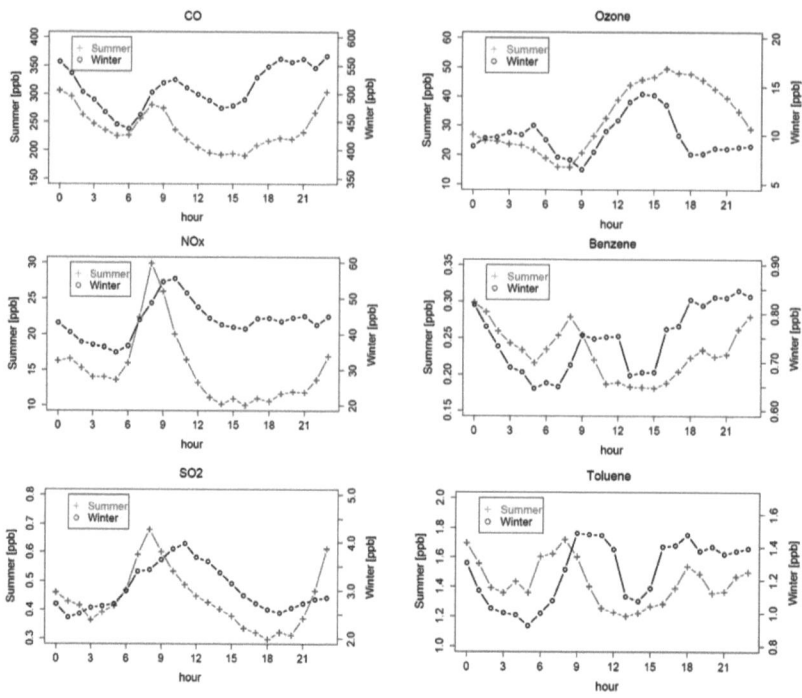

Figure 4. Average diurnal cycles of CO, NOx, SO$_2$, benzene, toluene and ozone in Zürich during the summer and winter campaigns. (Note different scales for summer and winter.)

Fig. 5 shows the diurnal variations for isoprene, MVK, methanol, acrolein, iso-propanol and MTBE in summer and in winter. All these compounds were identified in traffic exhaust emissions measured recently in a road tunnel close to Zürich (Legreid et al., 2007b). As for the trace gases discussed in Fig. 4 the late evening maximum for all compounds during cold start and from evaporation. The summer values of isoprene are most probably explained by biogenic emissions as reflected in the diurnal cycle. Temperature and photosynthetically active radiation (PAR) are the main factors controlling the emissions of isoprene (Guenther et al., 1993). However, in urban environments a part of the emissions could also be due to traffic exhaust emissions (Reimann et al., 2000; Borbon et al., 2004). In fact, by closer inspection of the isoprene diurnal cycle, two small peaks can be seen at rush traffic hours at 8 AM and 4PM.

5. OVOC measurements in Zürich

MVK is a product from the photooxidation of isoprene and has a local maximum in summer around noon. The relative increase of the MVK mixing ratio from the morning low to the daytime high is about 50% in summer. In winter, the MVK mixing ratio follows the CO mixing ratio quite closely, suggesting combustion sources for this compound. This has also been observed in other studies (Schauer et al., 1999, 2001, 2002). In contrast to the traffic related primary air pollutants, methanol shows no distinct late morning maximum. This suggests that even in an urban environment biogenic sources for methanol are dominant, which is underlined by the much higher mixing ratios in summer. From other studies methanol have also been reported to originate from biomass burning, both directly and secondarily from tropospheric oxidation processes (Holzinger et al., 1999; Holzinger et al., 2005).

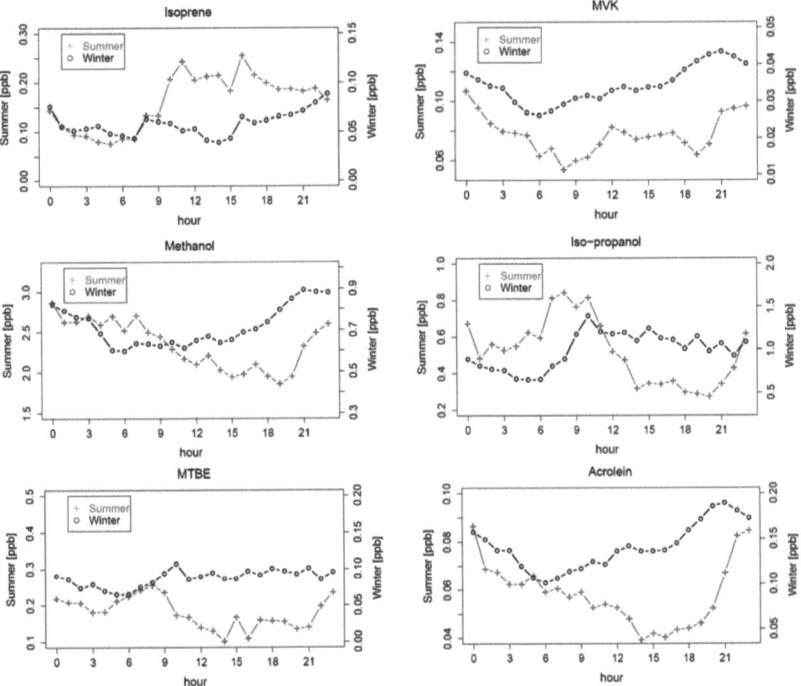

Figure 5. Average diurnal cycles of isoprene, MVK, methanol, acrolein, iso-propanol and MTBE in Zürich during the summer and winter campaigns (Note the different scales for summer and winter).

3.3 Source identification using a source-tracer-ratio method

3.3.1 Summer campaign

The results of the source allocation for the summer campaign using the source-tracer-ratio method described in Section 2.2 are shown in Table 2. Acetaldehyde, butanal, pentanal and acetone were the compounds with the highest relative contribution from the large-scale background (see Table 2), which can be explained by their secondary production from the oxidation of NMHCs (Atkinson, 2000). In fact, for acetaldehyde the large-scale background (41%) was the most important contribution to its mixing ratio in Zürich. This compound has been measured throughout the atmosphere and has been found even in the Arctic at levels over 200 ppt (Boudries et al., 2002). At other remote locations in the northern hemisphere mixing ratios of 200–400 ppt have been observed (Wisthaler et al., 2002; Singh et al., 2004). The main portion of the other aldehydes was explained by the regional background. The regional background probably includes more distant emissions as well as secondary production. Combustion explains less than 30% of the aldehydes mixing ratio in summer. The unsaturated aldehydes acrolein and methacrolein showed a different behaviour with negligible large-scale background due to their relatively low lifetime of less than a day. Acrolein was the aldehyde with the highest relative emission from combustion, whereas methacrolein had an important contribution from other sources, which in this case probably was dominated by secondary production from the oxidation of isoprene.

Apart from direct emissions, ketones (i.e. acetone, MEK, MVK) can also be produced by secondary formation in the atmosphere (Atkinson, 2000). Furthermore, the atmospheric lifetime of acetone is about 1 month. The combination of these two effects explains the relatively large amount of acetone in the large-scale background. Relatively high mixing ratios of acetone (0.5–1.1 ppb) have also been observed by other groups measuring this compound at remote locations (Boudries et al., 2002; Wisthaler et al., 2002; Holzinger et al., 2005; Schade and Goldstein, 2006). In summer large sources from vegetation have been reported for acetone (Goldstein and Schade, 2000; Jacob et al., 2002; Fall, 2003). In our study, these were incorporated in both the large-scale and the regional background, which explained in total 78% of the measured acetone mixing ratio in summer. MVK is a degradation product from isoprene, and MEK is an industrial solvent and also a secondary trace gas. Both compounds had a low large-scale background in this study, but could to a greater extent be attributed to the regional background. For MVK this can be explained by high regional atmospheric production in combination with the relatively low atmospheric lifetime of less than a day. MEK has an atmospheric lifetime of about 10 days, but low atmospheric production. Its regional background is probably explained by industrial sources due to its use as solvent in the manufacturing industry (EPA, 1994). MTBE could mainly be attributed to combustion due to its use as a gasoline additive in Switzerland. This compound is toxic and water soluble, which makes it a threat to drinking water sources.

Alcohols had a small relative contribution from the large-scale background except for methanol with 11%. Methanol has a long lifetime of 12 days and has consequently been found in relatively high quantities (0.3–1.2 ppb) at distant locations (Singh et al., 2000; Boudries et al., 2002; Wisthaler et al., 2002; Singh et al., 2004). It has been reported to mainly originate from biogenic sources (Jacob et al., 2005; Schade and Goldstein, 2006), but also from secondary production in altered biomass burning plumes (Holzinger et al., 2005). The other alcohols, except n-butanol, could mainly be explained by combustion processes, either from vehicle or industrial emissions. The high relative contribution of the combustion source to ethanol and iso-propanol can be explained by their usage in window wiper fluids. For n-butanol industrial sources were shown to be dominant in winter (see below), which in summer were included into the fraction of "combustion"and "other sources."The esters (methyl acetate, ethyl acetate and butyl acetate) were mainly attributed to the regional background. These compounds are common solvents in the industry, and ethyl acetate and butyl acetate are also used in industrial food production. In addition, methyl acetate has a considerable contribution from the large-scale background, which can be explained by its stability with a lifetime of about 30 days due to the reaction with OH radical. It is produced in the atmosphere from the oxidation of MTBE (Smith et al., 1991; Tuazon et al., 1991) and tert-amyl methyl ether (TAME) (both gasoline additives, Smith et al., 1995), and is also found in air from landfills (Fernandez-Martinez et al., 2001).

5. OVOC measurements in Zürich

	Local sources		Background	
	Combustion (%)	Other sources (%)	Regional (%)	Large scale (%)
Formaldehyde	12	15	60	13
Acetaldehyde	19	16	25	41
Propanal	19	12	56	13
Butanal	12	12	49	26
Pentanal	18	13	42	26
Hexanal	29	11	49	11
Benzaldehyde	31	9	50	10
Acrolein	34	9	52	4
Methacrolein	16	20	61	2
Methyl-tert-butyl-ether	50	6	43	2
Acetone	9	14	48	30
MVK	12	26	55	6
MEK	20	13	59	8
Methanol	18	8	63	11
Ethanol	55	11	32	1
Isopropanol	43	17	37	3
Propanol	41	23	33	4
2-Methyl-3-buten-2-ol	50	8	39	3
Butanol	17	69	14	NA
Methyl acetate	28	7	48	16
Ethyl acetate	35	23	40	2
Butyl acetate	35	26	36	3

Table 2. Source allocation for OVOCs in summer. The local sources describe the daily accumulation (see Section 3.2).

In Table 3 the contributions of the background and regional sources to NMHC concentrations in summer in Zürich are shown. Generally, the NMHCs have a low contribution from the largescale background (0–5%), and again the compound with longest lifetime (benzene) had the highest contribution. Most of the NMHCs were mainly explained by regional background, which could originate from distant road traffic and industrial sources. The exceptions were 1,3-butadiene and isoprene due to their low lifetimes of only a few hours in summer. In the case of 1,3-butadiene local industry could be important, which was also suspected from a study in France (Borbon et al., 2004). The high fraction of other sources explaining the isoprene emissions were most likely of biogenic origin as estimated for European conditions by Simpson et al. (1995).

5. OVOC measurements in Zürich

	Local sources		Background	
	Combustion (%)	Other sources (%)	Regional (%)	Large scale (%)
Butane	34	6	58	2
1,3-Butadiene	26	55	18	1
Isoprene	8	66	26	0
Benzene	31	5	59	5
Toluene	23	16	60	1
Ethylbenzene	25	9	64	1
m,p-Xylene	31	8	61	1
o-Xylene	29	10	60	1
1,2,4-Trimethylbenzene	35	27	37	1

Table 3. Source allocation for NMHCs in summer.

Fig. 6 shows the summary of the source profiles for the OVOCs and NMHCs in summer in Zürich obtained from the source-tracer-ratio method. The OVOCs compared to the NMHCs due to secondary production of the important compounds acetaldehyde, acetone and methanol. The regional background was the largest contributor for both groups of compounds with 50% and 56% contribution to the OVOCs and NMHCs, respectively. The main local source was combustion, which explained about one-quarter of the mixing ratios for both groups of compounds. Other sources in summer could be emissions from biogenic sources, from industry (explained in winter 4% of the OVOC emissions) and households.

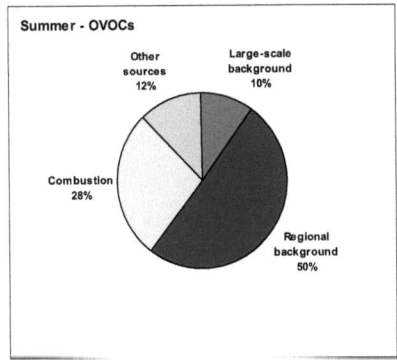
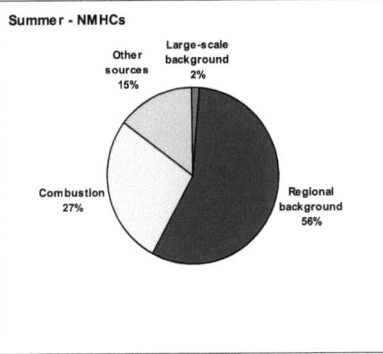

Figure 6. Source profile for OVOCs and NMHCs in Zürich in summer. (Other sources mean local biogenic sources, solvent use or locally produced secondary compounds).

3.3.2 Winter campaign

The relative contributions to the OVOC mixing ratios from the various sources in winter are shown in Table 4. For the majority of the aldehydes, the relative contribution from the large-scale background was higher in winter than in summer. This might reflect the longer atmospheric lifetimes in this season. Lower biogenic sources were the most likely reason for the reduced contribution of the regional background for acetaldehyde, butanal and hexanal. The combustion source contribution increased for all aldehydes in winter compared to summer, which is probably due to elevated emissions from stationary combustion sources like oil burners and stoves. The fuel for these units are similar to diesel, and diesel-fuelled vehicles are known to emit more oxygenated compounds than gasoline-fuelled vehicles per litre of fuel (Schauer et al., 1999, 2002; Legreid et al., 2007b).

MTBE has the same contribution from combustion sources in winter as in summer since it is solely emitted from gasoline vehicles. In winter the ketones have a higher contribution from the large-scale background than from the regional background. This might be explained by the increased lifetime, and in the case of acetone and MVK the reduced contribution of the regional background is due to lower emissions from biogenic sources (acetone) and regional secondary production (MVK and acetone) in wintertime. The combustion sources for these two compounds were elevated in winter for the same reason as for the aldehydes. MEK had an additional contribution from industrial sources, reflecting its use as an industrial solvent (EPA, 1994).

With the exception of methanol, a negligible contribution from the large-scale background was observed for the alcohols in both winter and summer. Methanol has a much higher relative contribution from the large-scale background due to the longer lifetime and lower contribution of regional biogenic emissions during the winter. The other alcohols have a somewhat higher contribution from the regional background and lower influence from the combustion sources except for n-butanol, for which almost half of the mixing ratio was explained by a local industry source in winter. Methyl acetate has a very similar source pattern in winter as in summer. Ethyl and butyl acetate had a distinct weekday/weekend difference in the ratio to CO, which was explained by local industrial sources like printing shop and small industries (Niedojadlo, 2001).

5. OVOC measurements in Zürich

	Combustion (%)	Industry (%)	Other sources (%)	Regional background (%)	Large-scale background (%)
Formaldehyde	19		1	63	7
Acetaldehyde	36		6	4	55
Propanal	26		5	53	16
Butanal	29		5	6	61
Pentanal	39			49	12
Hexanal	36			28	36
Benzaldehyde	43			41	17
Acrolein	41		7	45	7
Methacrolein	49		0	42	9
Methyl-tert-butyl-ether	50			46	4
Acetone	16	16		17	52
MVK	29		8	46	16
MEK	15	18		49	18
Methanol	26		1	-4	65
Ethanol	41		1	44	1
Isopropanol	38	17		42	3
Propanol	13	32		53	2
2-Methyl-3-buten-2-ol	48			49	3
Butanol	30	14		57	NA
Methyl acetate	27		1	42	21
Ethyl acetate	16	34		45	5
Butyl acetate	9	32		57	2

Table 4. Source allocation for OVOCs in winter.

Table 5 shows the estimated NMHC contributions to the measured mixing ratios. Both butane and benzene have a higher fraction explained by the large-scale background in winter than in summer, due to the longer atmospheric lifetime in this season. The more reactive NMHCs have negligible amounts in the large-scale background, and are mainly explained by the regional background. Combustion has a larger relative importance in winter than in summer due to more intensive stationary combustion and cold-start emissions from mobile sources.

5. OVOC measurements in Zürich

	Local sources			Background	
	Combustion (%)	Industry	Other sources (%)	Regional (%)	Large scale (%)
Butane	3		3	53	9
1,3-Butadiene	4		12	45	2
Isoprene	4		22	32	0
Benzene	3		4	52	13
Toluene	4			52	1
Ethylbenzene	3		6	60	1
m,p-Xylene	4			51	1
o-Xylene	4			53	1
1,2,4-Trimethylbenzene	3			58	3

Table 5. Source allocation for NMHCs in winter.

Fig. 7 shows the summary of the source profiles for the OVOCs and NMHCs in Zürich in winter. Compared to summer, the contribution of both the regional and large-scale background was lower for the OVOCs, whereas the contribution of the combustion is elevated. This is due to lesser biogenic emissions and secondary production combined with higher emissions from stationary and mobile combustion. A similar pattern is found for the NMHCs, but for these compounds the large-scale background is elevated compared to summer.

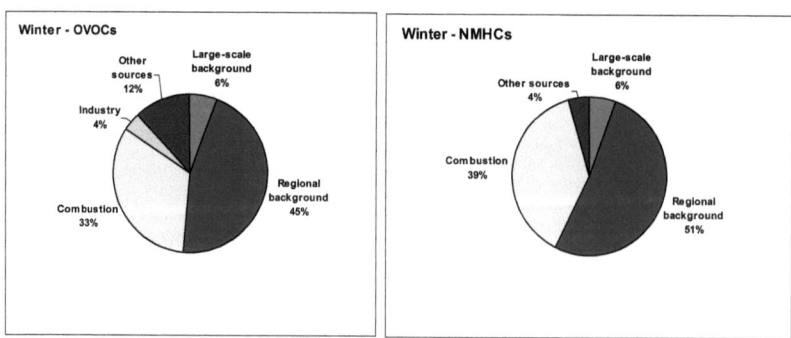

Figure 7. Source profile for OVOCs and NMHCs in Zürich in winter

4. Conclusions

Four seasonal campaigns where OVOCs and selected NMHCs have been measured in Zürich have been presented in this paper. The collected data are unique since they represent the first dataset with a large number of OVOCs from one European urban site during all seasons. Ethanol was the compound with the highest mixing ratios at all seasons. There was a clear annual cycle for compounds known to be emitted from biogenic sources in summer. This resulted in more than two times higher mixing ratios for methanol, isoprene, hex-anal and acetone in summer compared to winter, which implies that biogenic sources are dominant for these compounds. Oxidation products of isoprene were also elevated in summer due to atmospheric production. This was the case for methacrolein and MVK which both had two times higher mixing ratios in summer. Other compounds like ethanol, propanol, acrolein and benzene showed an opposite trend due to more intensive stationary combustion in winter. By the sourcetracer-ratio method, combustion was estimated to be the main local source during both seasons, with 28% in summer and 33% in winter. For the measured NMHCs, combustion explained 27% and 39%, respectively. Overall, the main contributions to measured concentrations were explained by the regional and large-scale background.

5. Acknowledgements

This study was financially supported by the Swiss Federal Office for the Environment (FOEN/BAFU). For helpful discussions we thank P.G. Simmonds and D. Young at the University of Bristol. We also thank MeteoSwiss for the preparation of meteorological data and the NABEL team for supporting the maintenance work at the measurement station.

References:

Altshuller, A. P. (1978). "Assessment Of Contribution Of Chemical Species To Eye Irritation Potential Of Photo-Chemical Smog." Journal Of The Air Pollution Control Association **28**(6): 594-598.

Apel, E. C., D. D. Riemer, et al. (2002). "Measurement and interpretation of isoprene fluxes and isoprene, methacrolein, and methyl vinyl ketone mixing ratios at the PROPHET site during the 1998 Intensive." Journal of Geophysical Research-Atmospheres **107**(D3): art. no.-4034.

Atkinson, R. (2000). "Atmospheric chemistry of VOCs and NOx." Atmospheric Environment **34**(12-14): 2063-2101.

Bakeas, E. B., D. I. Argyris, et al. (2003). "Carbonyl compounds in the urban environment of Athens, Greece." Chemosphere **52**(5): 805-813.

Balzani, J., 2007. Carbonyls and PANs at Jungfraujoch and the related oxidation processes at the boundary layer/free troposphere interface, Ph.D. Thesis Nr. 17234, ETH Zurich, Switzerland.

Borbon, A., P. Coddeville, et al. (2004). "Characterising sources and sinks of rural VOC in eastern France." Chemosphere **57**(8): 931-942.

Boudries, H., J. W. Bottenheim, et al. (2002). "Distribution and trends of oxygenated hydrocarbons in the high Arctic derived from measurements in the atmospheric boundary layer and interstitial snow air during the ALERT2000 field campaign." Atmospheric Environment **36**(15-16): 2573-2583.

EMEP (2005). "Expert Emissions used in EMEP models." http://webdab.emep.int.

EPA (1994). "Locating and estimating air emissions from sources of methyl ethyl ketone."

Fall, R. (2003). "Abundant oxygenates in the atmosphere: A biochemical perspective." Chemical Reviews **103**(12): 4941-4951.

Fernandez-Martinez, G., P. Lopez-Mahia, et al. (2001). "Measurement of volatile organic compounds in urban air of La Coruna, Spain." Water Air And Soil Pollution **129**(1-4): 267-288.

Goldan, P. D., M. Trainer, et al. (1995). "Measurements Of Hydrocarbons, Oxygenated Hydrocarbons, Carbon-Monoxide, And Nitrogen-Oxides In An Urban Basin In Colorado - Implications For Emission Inventories." Journal Of Geophysical Research-Atmospheres **100**(D11): 22771-22783.

Goldstein, A. H. and G. W. Schade (2000). "Quantifying biogenic and anthropogenic contributions to acetone mixing ratios in a rural environment." Atmospheric Environment **34**(29-30): 4997-5006.

Granby, K., C. S. Christensen, et al. (1997). "Urban and semi-rural observations of carboxylic acids and carbonyls." Atmospheric Environment **31**(10): 1403-1415.

Guenther, A. B., P. R. Zimmerman, et al. (1993). "Isoprene And Monoterpene Emission Rate Variability - Model Evaluations And Sensitivity Analyses." Journal Of Geophysical Research-Atmospheres **98**(D7): 12609-12617.

Hak, C., I. Pundt, et al. (2005). "Intercomparison of four different in-situ techniques for ambient formaldehyde measurements in urban air." Atmospheric Chemistry And Physics **5**: 2881-2900.

Heck, W. W., W. W. Cure, et al. (1984). "Assessing Impacts Of Ozone On Agricultural Crops .1. Overview." Journal Of The Air Pollution Control Association **34**(7): 729-735.

Hewitt, C. N. (1999). "Reactive Hydrocarbons in the Atmosphere."

Holzinger, R., C. Warneke, et al. (1999). "Biomass burning as a source of formaldehyde, acetaldehyde, methanol, acetone, acetonitrile, and hydrogen cyanide." Geophysical Research Letters **26**(8): 1161-1164.

Holzinger, R., J. Williams, et al. (2005). "Oxygenated compounds in aged biomass burning plumes over the Eastern Mediterranean: evidence for strong secondary production of methanol and acetone." Atmospheric Chemistry And Physics **5**: 39-46.

Jacob, D. J., B. D. Field, et al. (2002). "Atmospheric budget of acetone." Journal Of Geophysical Research-Atmospheres **107**(D10).

Jacob, D. J., B. D. Field, et al. (2005). "Global budget of methanol: Constraints from atmospheric observations." Journal Of Geophysical Research-Atmospheres **110**(D8).

Jonsson, A., K. A. Persson, et al. (1985). "Measurements Of Some Low Molecular-Weight Oxygenated, Aromatic, And Chlorinated Hydrocarbons In Ambient Air And In Vehicle Emissions." Environment International **11**(2-4): 383-392.

Junkermann, W. and J. M. Burger (2006). "A new portable instrument for continuous measurement of formaldehyde in ambient air." Journal Of Atmospheric And Oceanic Technology **23**(1): 38-45.

Kim, E., S. G. Brown, et al. (2005). "Characterization of non-methane volatile organic compounds sources in Houston during 2001 using positive matrix factorization " ATMOSPHERIC ENVIRONMENT **39**(32): 5934-5946.

Komenda, M., A. Schaub, et al. (2003). "Description and characterization of an on-line system for long- term measurements of isoprene, methyl vinyl ketone, and methacrolein in ambient air." Journal of Chromatography A **995**(1-2): 185-201.

Kuhlwein, J., R. Friedrich, et al. (2002). "Comparison of modelled and measured total CO and NOx emission rates." Atmospheric Environment **36**: S53-S60.

Legreid, G., D. Folini, et al. (2008). "Measurements of organic trace gases including OVOCs at the high alpine site Jungfraujoch (Switzerland): Seasonal variation and source allocations." Journal Of Geophysical Research-Atmospheres **113** (D05307).

Legreid, G., S. Reimann, et al. (2007). "Measurements of OVOCs and NMHCs in a Swiss highway tunnel for estimation of road transport emissions." Environmental Science & Technology **41**: 7060-7066.

McDonald, R. D., B. Zielinska, et al. (2000). "Fine particle and gaseous emission rates from residential wood combustion." Environmental Science & Technology **34**(11): 2080-2091.

Millet, D. B., N. M. Donahue, et al. (2005). "Atmospheric volatile organic compound measurements during the Pittsburgh Air Quality Study: Results, interpretation, and quantification of primary and secondary contributions." Journal Of Geophysical Research-Atmospheres **110**(D7).

NABEL (2004). "NABEL - Luftbelastung 2004. Schriftenreihe Umwelt Nr. 388, BUWAL 2005." 30-32.

Nguyen, H. T. H., N. Takenaka, et al. (2001). "Atmospheric alcohols and aldehydes concentrations measured in Osaka, Japan and in Sao Paulo, Brazil." Atmospheric Environment **35**(18): 3075-3083.

Niedojadlo, A. (2001). "Measurements of Oxygenated VOC Emissions from Solvent Use in the City Air of Wuppertal." Extended Abstract at Conference: "Oxygenated Organics in the Atmosphere, Sources, Sinks and Atmospheric Impact ".

Northway, M. J., J. A. de Gouw, et al. (2004). "Evaluation of the role of heterogeneous oxidation of alkenes in the detection of atmospheric acetaldehyde." Atmospheric Environment **38**(35): 6017-6028.

Odum, J. R., T. Hoffmann, et al. (1996). "Gas/particle partitioning and secondary organic aerosol yields." Environmental Science & Technology **30**(8): 2580-2585.

Odum, J. R., T. P. W. Jungkamp, et al. (1997). "Aromatics, reformulated gasoline, and atmospheric organic aerosol formation." Environmental Science & Technology **31**(7): 1890-1897.

Placet, M., C. O. Mann, et al. (2000). "Emissions of ozone precursors from stationary sources: a critical review." Atmospheric Environment **34**(12-14): 2183-2204.

Possanzini, M., V. Dipalo, et al. (1996). "Measurements of lower carbonyls in Rome ambient air." Atmospheric Environment **30**(22): 3757-3764.

Poulopoulos, S. and C. Philippopoulos (2000). "Influence of MTBE addition into gasoline on automotive exhaust emissions." Atmospheric Environment **34**(28): 4781-4786.

Reimann, S., P. Calanca, et al. (2000). "The anthropogenic contribution to isoprene concentrations in a rural atmosphere." Atmospheric Environment **34**(1): 109-115.

Sawyer, R. F., R. A. Harley, et al. (2000). "Mobile sources critical review: 1998 NARSTO assessment." Atmospheric Environment **34**(12-14): 2161-2181.

Schade, G. W., G. B. Dreyfus, et al. (2002). "Atmospheric methyl tertiary butyl ether (MTBE) at a rural mountain site in California." Journal Of Environmental Quality **31**(4): 1088-1094.

Schade, G. W. and A. H. Goldstein (2006). "Seasonal measurements of acetone and methanol: Abundances and implications for atmospheric budgets." Global Biogeochemical Cycles **20**(1).

Schauer, J. J., M. J. Kleeman, et al. (1999). "Measurement of emissions from air pollution sources. 2. C-1 through C-30 organic compounds from medium duty diesel trucks." Environmental Science & Technology **33**(10): 1578-1587.

Schauer, J. J., M. J. Kleeman, et al. (2001). "Measurement of emissions from air pollution sources. 3. C-1-C-29 organic compounds from fireplace combustion of wood." Environmental Science & Technology **35**(9): 1716-1728.

Schauer, J. J., M. J. Kleeman, et al. (2002). "Measurement of emissions from air pollution sources. 5. C-1-C-32 organic compounds from gasoline-powered motor vehicles." Environmental Science & Technology **36**(6): 1169-1180.

Simpson, D., A. Guenther, et al. (1995). "Biogenic Emissions In Europe .1. Estimates And Uncertainties." Journal Of Geophysical Research-Atmospheres **100**(D11): 22875-22890.

Singh, H., Y. Chen, et al. (2000). "Distribution and fate of selected oxygenated organic species in the troposphere and lower stratosphere over the Atlantic." Journal Of Geophysical Research-Atmospheres **105**(D3): 3795-3805.

Singh, H. B., L. J. Salas, et al. (2004). "Analysis of the atmospheric distribution, sources, and sinks of oxygenated volatile organic chemicals based on measurements over the Pacific during TRACE-P." Journal of Geophysical Research-Atmospheres **109**(D15): art. no.-D15S07.

Smith, D. F., T. E. Kleindienst, et al. (1991). "The Photooxidation Of Methyl Tertiary Butyl Ether." International Journal Of Chemical Kinetics **23**(10): 907-924.

Smith, D. F., C. D. McIver, et al. (1995). "Kinetics And Mechanism Of The Atmospheric Oxidation Of Tertiary Amyl Methyl-Ether." International Journal Of Chemical Kinetics **27**(5): 453-472.

Staehelin, J., R. Locher, et al. (2001). "Contribution of road traffic emissions to ambient air concentrations of hydrocarbons: the interpretation of monitoring measurements in

Switzerland by Principal Component Analysis and road tunnel measurements." International Journal Of Vehicle Design **27**(1-4): 161-172.

Steinbacher, M.J., Dommen, J., Ordonez, C., Reimann, S., Grüebler, F.C., Staehelin, J., Prevot, A.S.H., 2005. Volatile organic compounds in the Po Basin. Part A: anthropogenic VOCs. Journal of Atmospheric Chemistry **51**: 271–291.

Tuazon, E. C., W. P. L. Carter, et al. (1991). "Products Of The Gas-Phase Reaction Of Methyl Tert-Butyl Ether With The Oh Radical In The Presence Of Nox." International Journal Of Chemical Kinetics **23**(11): 1003-1015.

Wang, D., J. D. Fuentes, et al. (2005). "Non-methane hydrocarbons and carbonyls in the Lower Fraser Valley during PACIFIC 2001." Atmospheric Environment **39**(29): 5261-5272.

Wisthaler, A., A. Hansel, et al. (2002). "Organic trace gas measurements by PTR-MS during INDOEX 1999." Journal of Geophysical Research-Atmospheres **107**(D19): art. no.-8024.

Zhao, W. X., P. K. Hopke, et al. (2004). "Source identification of volatile organic compounds in Houston, Texas." Environmental Science & Technology **38**(5): 1338-1347.

6 OVOC measurements at Jungfraujoch

Legreid, G., D. Folini, et al., "Measurements of organic trace gases including OVOCs at the high alpine site Jungfraujoch (Switzerland): Seasonal variation and source allocations." Journal Of Geophysical Research-Atmospheres **113** (D05307), doi: 10.1029/2007JD008653, 2008. Copyright 2012 American Geophysical Union.

Reproduced by permission of American Geophysical Union

Measurements of organic trace gases including OVOCs at the high alpine site Jungfraujoch (Switzerland): Seasonal variation and source allocations

Geir Legreid[†], Doris Folini[†], Johannes Staehelin[‡], Jacob Balzani Lööv[‡], Martin Steinbacher[†] and Stefan Reimann[†]*

[†] Empa, Swiss Federal Laboratories for Materials Testing and Research, Laboratory for Air Pollution and Environmental Technology, Ueberlandstrasse 129, CH-8600 Duebendorf, Switzerland

[‡] Institute for Atmospheric and Climate Science, Swiss Federal Institute of Technology, CH-8092 Zurich, Switzerland,

*Corresponding author phone: +41-44-823-5511; fax: +41-44-621-6244; e-mail: Stefan.reimann@empa.ch

6. OVOC measurements at Jungfraujoch

Abstract

At the high alpine site Jungfraujoch (Switzerland) mixing ratios of 21 oxygenated volatile organic compounds (OVOCs) and selected non-methane hydrocarbons (NMHCs) have been measured by a newly developed two-stage double adsorbent system coupled to gas chromatograph-mass spectrometer (GC-MS). In addition, formaldehyde was measured by the Hantzsch technique. Four measurement campaigns were performed once every season in 2005, providing for the first time a unique data set of OVOCs in the free troposphere of central Europe. The dominating OVOCs measured were acetone, methanol, formaldehyde and acetaldehyde, with mean mixing ratios of 622–867 ppt, 362–790 ppt, 303–505 ppt and 310–392 ppt, respectively. These compounds explained 95% of the measured organic compounds in summer and 83% in fall. Elevated mixing ratios in summer were observed for compounds with strong biogenic sources (e.g., methanol and acetone), whereas mainly anthropogenic compounds (e.g., ethanol and benzene) had higher mixing ratios during winter. Potential European source regions were estimated for the organic trace gases by combining the measured data with a statistical trajectory model. Northern Italy, southern France and southern and eastern Germany were identified as the main European contributors to the measured organic compounds at Jungfraujoch.

1   Introduction

The high alpine site Jungfraujoch (Switzerland) is located in central Europe at 3580 m asl, and is an excellent location for studying atmospheric chemical processes related to the European boundary layer and the background troposphere [Reimann et al., 2004; Zellweger et al., 2003]. Oxygenated volatile organic compounds (OVOCs) play a key role in tropospheric photochemistry as both precursors and intermediates in tropospheric production of ozone and other photo oxidants. [Atkinson, 2000; Wennberg et al., 1998]. OVOCs originate from both direct emissions by biogenic and anthropogenic sources, and secondary production from the oxidation of methane and nonmethane hydrocarbons (NMHCs). Apart from weekly measurements of C1–C4 aldehydes at 5 remote rural European stations since 1992/1993 [Solberg et al., 1996] and measurements of acetone and methanol at a rural site in USA during 1 year [Schade and Goldstein, 2006], measurements of OVOCs in the background troposphere have only been reported from few temporally limited campaigns. In summer 2002, Lewis et al. [2005] reported acetone, methanol, acetaldehyde and several NMHC mixing ratios in North Atlantic air, and found that the three OVOCs explained up to 85% of the total NMHCs mass under maritime conditions. Singh et al. [2000] measured selected OVOCs in the troposphere and lower stratosphere over the Atlantic during an aircraft campaign and concluded that OVOCs are important sources of atmospheric free radicals and are also linked to free tropospheric ozone formation. OVOCs were also measured during the ALERT2000 field campaign in the high Artic, where unexpectedly high mixing ratios of acetone, methanol and acetaldehyde were found [Boudries et al., 2002]. These compounds accounted for about 90% of the OVOCs in the Arctic troposphere, with photochemical production in the snow layer as a suspected main source. Furthermore, Wisthaler et al. [2002] measured methanol, acetaldehyde and acetone with proton reaction transfer–mass spectrometer (PTR-MS) over the Indian Ocean during INDOEX in 1999. High mixing ratios of methanol were discovered in biomass burning plumes, whereas acetaldehyde and acetone were elevated in air masses with high anthropogenic input. In addition, OVOCs were measured in marine boundary layer air at Trinidad Head, California [Millet et al., 2004], and height profiles of several OVOCs were reported from measurements over the Pacific in 2001 [Singh et al., 2004]. In the Alps measurements have been performed at the alpine background site of Sonnblick (Austria) [Karl et al., 2001a, 2001b]. In the higher troposphere over Europe Colomb et al. [2006] presented measurements of formaldehyde, methanol, acetone and toluene at 6–13 km altitude in summer. At 6 km altitude the mean mixing ratios reported were about 100 ppt for formaldehyde, 800 ppt for methanol, 1300 ppt for acetone and 50 ppt for toluene.

In this study continuous state-of-the-art measurements of 21 OVOCs and selected NMHCs are presented from a remote location in Europe, performed during 4 seasonal campaigns lasting about 1 month each. In the following section a brief description of the instrument for measuring the OVOCs is given. This is followed by an explanation of the applied technique for separation of air masses into background conditions and periods when the station was influenced by the polluted boundary layer. In addition, the method for statistical trajectory analysis used for the source allocation is summarized. In section 3 the seasonal variations are discussed and case studies are presented, which show characteristics for the free tropospheric conditions at this high-altitude central European site. Furthermore, European sources are allocated by analyzing the measurements with trajectory analysis.

2 Measurements and methods

2.1 Trace gas measurements

The high alpine station Jungfraujoch (46°33'N, 7°59'E, 3580 m asl) is located on a mountain saddle between the Jungfrau (4158 m asl) in the west and Moench (4099 m asl) in the east. The sampling point located at the top of the mountain ridge was used for the measurements of OVOCs, NMHCs and other trace gases measured by the Swiss Network for Air Pollution Monitoring (NABEL). From the heated inlet (10°C) a ¼" x 12 m PFA line was used as connection to the main sampling pump of the instrument, which was run at a flow of 500 ml/min. In order to avoid artefact

6. OVOC measurements at Jungfraujoch

formation of aldehydes by ozone during sampling, NO-titration was used in the summer and autumn campaigns for ozone destruction [Komenda, et al., 2003] with 10 ml/min of 20 ppm NO in N_2. The produced NO_2 can be seen as non-reactive on the time scale of 0.5 minutes (i.e. the time which the air was in the sampling line). From the main sampling line a ¼" x 0.5 m PFA line was connected to the sampling unit. The flow from this line was kept at 40 ml/min and the sample was collected during 12 minutes, which resulted in a total sample volume of 480 ml. The frequency was about one sample every 50 minutes.

The air samples were collected on a two-stage adsorbent system connected to a gas chromatograph-mass spectrometer (GC-MS Agilent HP 6890 – HP 5973N). In short, the analytes were sampled on a mainly hydrophobic adsorbent 0.6 g Hayesep-D trap at room temperature. By this procedure most of the water was not trapped. The rest humidity was removed from the adsorbent by dry helium flushing. The analytes were then released by heating the trap to 160 °C and they were refocused on a 14 mg Hayesep D trap (- 40 °C) to ensure good separation on the GC column. For further details, see Legreid et al. [Legreid, et al., 2006]. The OVOC were measured with an accuracy of 3–25 % (n-butanol: 37 %) and a precision of 1–5 %, calculated from intercomparison experiments. This intercomparison was a part of the European project ACCENT, subgroup Quality Assurance (QA). Methanol was only recovered at 60 %, which was corrected for in the measurement campaigns. The detection limit for each compound was calculated as 3 times the standard deviation above noise of 5 zero air samples. Calibration of the instrument was performed once per day by analysis of a standard gas mixture (Apel & Riemer Environmental Inc., USA).

The OVOC system was in operation during four measurement campaigns in winter, spring, summer and autumn 2005. The winter measurements were performed from 8th of February until 8th of March 2005 during CLACE-4 (Cloud and Aerosol Characterisation Experiment in the Free Troposphere), a campaign which was focused on ice cloud formation. Spring measurements followed from 22nd of April until 30th of May, summer measurements from 5th of August until 19th of September and fall measurements from 14th of October until 1st of November. A total of about 2'400 samples were analyzed

Measurements of other traces at Jungfraujoch are continuously performed by NABEL. NMHCs (n-pentane, iso-pentane, n-hexane, benzene and toluene) were measured with a GC-MS system optimized for the analysis of CFCs and HFCs [Reimann, et al., 2004]. Nitrogen oxides (NO, NO2 and NOy) were monitored with chemiluminescence detectors, carbon monoxide (CO) with non-dispersive infrared technique (NDIR) and ozone (O3) with UV absorption [Zellweger, et al., 2000].

2.2 Filtering of data

Meteorological processes were recognized as being important for understanding the abundance of the OVOCs and NMHCs at the site. Therefore, measurement data were connected with information about the origin of air masses arriving at the station, also making use of trajectory analysis. The measurement data were first classified into either "undisturbed free tropospheric (FT) conditions" or "disturbed FT conditions". The term "disturbed FT" describes air masses that contain polluted boundary layer (PBL) air mixed with FT air during upward transport caused by different meteorological processes. Zellweger et al. [2003] estimated the contribution of PBL to undisturbed FT air masses to be 14-20% during convection events at Jungfraujoch. In the same paper it was suggested to use the NO_y/CO ratio to identify the undisturbed FT air masses. Recently, a similar approach was also applied by Zanis et al. [2006], when investigating the seasonal variability of the ozone production efficiencies in the lower free troposphere at Jungfraujoch, and by Walker et al. [2006], when studying the processes controlling the hydro peroxide concentrations at Jungfraujoch.

In this study the NO_y/CO ratio applied for the filter was 0.005 (winter), 0.006 (spring and fall) and 0.007 (summer). The air masses with lower ratios were classified as undisturbed FT. Disturbed FT conditions were characterized by NO_y/CO ratios higher than 0.007 (winter), 0.008 (spring and fall) and 0.009 (summer). For the source estimation the data with NO_y/CO ratio between 0.005 and 0.007 (winter), 0.006 and 0.008 (spring and fall) and 0.007 and 0.009 (summer) were ignored. For the identification of source regions, aLMo (alpine Model) 48-hours backward trajectories were applied, which are described in more detail in the following paragraph.

2.3 Source identification by backward trajectory model

Backward trajectories were calculated based on the wind fields from the operational numerical weather prediction model (alpine model, aLMo) of the Federal Office of Meteorology and Climatology of Switzerland (MeteoSwiss). The aLMo runs on a horizontal 7 km × 7 km grid and 45 vertical levels. Wind fields are available for each hour. 48 h backward trajectories are computed six times per day, every four hours. Their arrival point at Jungfraujoch is 100 m above model ground. To better account for complex flow regimes at the mountain site, additional four trajectories being displaced at the neighbouring grid points were used together with the trajectory arriving at the exact location. For the data analysis these trajectories were combined with the measured data in order to identify source locations for the compounds. In this study we used the method of Seibert et al. [1994].

The domain of the calculated trajectories was superimposed with a 0.22° × 0.22° grid. The mean mixing ratio for these grid cells were calculated after the approach of Seibert et al. [1994] by the following formula:

$$\overline{C_{mn}} = \frac{1}{\sum_{a=1}^{M} \tau_{mna}} \sum_{a=1}^{M} (c_a) \tau_{mna},$$

where $\overline{C_{mn}}$ is a relative measure of potential source region strength, m, n are indices of the horizontal grid, a is the index of the trajectory, M is the total number of trajectories, c_a is the concentration (minus the background concentration) measured during arrival of trajectory a, and τ_{mna} is the residence time (in units of trajectory time steps) that the trajectory spent within the PBL over grid cell m, n. We assume a trajectory to reside within the PBL if the pressure difference between the trajectory and surface is smaller than 120 hPa in winter and 200 hPa in summer.

A high $\overline{C_{mn}}$ value for a specific grid cell m,n means that over-passing air masses are on average associated with high mixing ratios at the receptor site. However, due to equal distributions of the measured mixing ratios to all grid cells passed by the appropriate trajectory, there is a possibility for underestimation of spatial gradients of true emissions [*Stohl*, 1996]. Other constraints are that by only measuring at one site the inhomogeneous advection regimes decrease the amount of independent information. The main influence region at Jungfraujoch is referred to as central western Europe (i.e. Switzerland, eastern France, southern Germany and northern Italy), which means that estimates from other parts of Europe are hindered by dilution processes during transport and low occurrence of transport from the respective area [*Reimann, et al.*, 2004].

3 Results and discussion

3.1 Description of data

The mixing ratios of the measured OVOCs and NMHCs during the four campaigns at Jungfraujoch in 2005 are presented in Table 1. The medians, means, interquartile ranges (25 – 75) (IQR) for each compound as well as the detection limits (3 σ above noise) are listed for each compound. Due to ozone interference during the winter and spring campaign, the values for the impacted carbonyls (e.g. acetaldehyde and propanal) are not shown. In the following the measurement data from Jungfraujoch are compared with those of other studies from remote sites (Table 2) and the results are discussed in the context of known sources.

6. OVOC measurements at Jungfraujoch

Compounds (All values in ppt)	Spring 2005 Median	Mean	IQR	Det.limit	Summer 2005 Median	Mean	IQR	Det.limit	Fall 2005 Median	Mean	IQR	Det.limit	Winter 2005 Median	Mean	IQR	Det.limit
Formaldehyde	377	397	266 - 494	na	466	505	359 - 574	na	293.45	303	250 - 350	na	268	362	171 - 537	na
Acetaldehyde	na	na	na	na	383	392	353 - 425	70	310	310	292 - 331	49	na	na	na	na
Propanal	na	na	na	na	24	25	20 - 28	11	20	21	17 - 24	8	na	na	na	na
Butanal	na	na	na	na	19	22	15 - 24	4	22.75	24	18 - 29	3	na	na	na	na
Pentanal	na	na	na	na	10	11	9 - 12	6	5	5	3 - 7	9	na	na	na	na
Hexanal	na	na	na	na	10	11	9 - 12	6	9	10	8 - 12	4	na	na	na	na
Benzaldehyde	9	9	8 - 10	2	4	4	3 - 6	20	8	8	7 - 9	12	3	4	2 - 5	5
Acrolein	15	16	13 - 17	6	5	5	4 - 7	5	2.5	3	2 - 3	2	11	12	8 - 15	9
Methacrolein	3	3	2 - 3	1	3	5	2 - 6	2	1	2	1 - 2	1	1	2	1 - 2	2
Methyl-t-butyl-ether	6	7	4 - 8	<1	6	8	4 - 10	<1	9	10	6 - 14	<1	7	8	4 - 12	<1
Acetone	749	774	578 - 902	24	862	867	718 - 991	30	640	667	576 - 724	47	560	622	487 - 749	46
MVK	8	10	6 - 12	9	10	13	7 - 16	5	4	4	3 - 5	1	7	7	5 - 9	10
MEK	56	59	45 - 69	1	35	37	24 - 47	3	34.5	39	28 - 45	1	83	101	63 - 135	2
Methanol	689	790	515 - 991	62	747	769	579 - 956	79	351.73	362	303 - 408	66	482	550	402 - 657	90
Ethanol	122	135	76 - 174	5	106	117	72 - 145	7	194	206	157 - 232	4	239	313	135 - 410	72
Isopropanol	24	28	19 - 32	6	na	na	na	na	na	na	na	na	48	62	34 - 76	13
Propanol	4	4	3 - 5	<1	2	2	1 - 2	5	3	4	3 - 4	na	7	10	4 - 11	3
2-Methyl-3-butene-2-ol	7	9	7 - 11	<1	5	6	4 - 7	<1	10	11	9 - 12	<1	14	18	12 - 24	<1
Methyl acetate	17	21	13 - 27	<1	13	13	11 - 15	<1	9	10	6 - 12	<1	17	22	11 - 31	<1
Ethyl acetate	13	18	9 - 21	<1	8	9	6 - 11	<1	4	5	3 - 5	<1	4	6	3 - 8	<1
Butyl acetate	4	4	3 - 5	5	4	5	3 - 6	<1	4	5	3 - 5	1	4	6	3 - 8	8
Butane	34	43	24 - 49	1	26	31	18 - 39	1	43	48	33 - 59	<1	187	209	132 - 252	<1
1,3-Butadiene	1	2	1 - 2	<1	1	1	1 - 1	<1	2	2	1 - 3	<1	1	2	1 - 4	<1
Isoprene	1	1	0 - 1	<1	13	46	8 - 49	<1	2	2	1 - 3	<1	2	2	1 - 2	<1
Benzene	23	25	17 - 29	1	16	18	12 - 20	1	20	21	16 - 24	<1	90	110	65 - 139	1
Toluene	59	70	42 - 81	9	20	27	12 - 33	<1	58	62	44 - 71	1	54	66	28 - 82	5
Ethylbenzene	16	18	12 - 22	1	1	2	1 - 2	<1	2	2	2 - 3	<1	8	10	4 - 13	2
m,p-Xylene	67	75	51 - 100	3	2	3	1 - 3	<1	5	6	4 - 7	<1	12	20	7 - 21	6
o-Xylene	5	7	3 - 8	1	1	1	1 - 2	<1	2	2	2 - 2	<1	5	8	3 - 10	3

Table 1. Results from four seasonal measurement campaigns at Jungfraujoch in 2005. The medians, means, IQRs (interquartile ranges, 25 – 75) and detection limits for the OVOCs and NMHCs are listed. na: not available (See text for further explanation).

6. OVOC measurements at Jungfraujoch

Compounds (All values in ppt)	High Arctic 2000: Mean ± SD	Trinidad Head, USA 2002: IQR	Pacific 2001: Mean ± SD	Indian Ocean 1999: Mean ± SD	Atlantic 1997: Range	Sonnblick 1999: Range
Formaldehyde			188 ± 133			
Acetaldehyde	166 ± 79		226 ± 89	350 ± 32		
Propanal	11 ± 5		77 ± 34			
Butanal		15–23				
Methacrolein		9–24				
Methyl-t-butyl-		1–6				
Acetone	871 ± 234	529–801	822 ± 295	1121 ± 72	630–920	1500–3500
2-Butenone		3–9				
Butanone	54 ± 44	45–76	75 ± 52			
Methanol	256 ± 135	611–1021	1250 ± 691	687 ± 106	400–800	800–2000
Ethanol	36 ± 31	75–168	77 ± 69		<50	
Iso-propanol		11–27				
2-Methyl-3-buten-		2–18				
Notes	Apr–May (24 h light period)	Apr–May	Feb–Apr 2–4 km	Mar (NHcX	Oct–Nov	fall/winter

Table 2. Overview over previous OVOC measurement campaigns performed in the high artic during ALERT 2000 [*Boudries, et al.*, 2002], at Trinidad Head (California) in 2002 [*Millet, et al.*, 2004], over the Pacific during TRACE-P in 2001 [*Singh, et al.*, 2004], over the Indian ocean during INDOEX 1999 [*Wisthaler, et al.*, 2002], over the Atlantic in 1997 [*Singh, et al.*, 2000] and at Sonnblick, Austria in 1999 [Karl et al., 2001a].

3.1.1 Methyl-tert-butyl-ether (MTBE) and NMHCs

MTBE and the aromatic NMHCs are atmospheric trace gases, which are exclusively emitted by anthropogenic sources. The average mixing ratio for MTBE was higher than from a background site in the USA (Trinidad Head) [*Millet, et al.*, 2004], which can be explained by the stronger influence of the polluted boundary layer at Jungfraujoch. Benzene was the most abundant NMHC with an IQR of 64-113 ppt in winter. In general, the NMHCs have higher mixing ratios in winter due to longer lifetime in this season. However, the NMHCs are present in much lower mixing ratios than the OVOCs with less than 10% of the total measured organic compounds in summer and fall and less than 20% in winter and spring (albeit fewer aldehydes were measured in the winter and spring campaigns).

3.1.2 Esters

To our knowledge, the first measurement results for esters at a background site are presented in this paper. Methyl and ethyl acetate were present at similar levels of about 5-30 ppt, whereas the mixing ratios of butyl acetate were mostly below 5 ppt. The esters are used as solvents in the industry, but can also be produced from the oxidation of alkenes with ozone [*Atkinson*, 2000]. They show high correlations during winter ($R^2 = 0.64–0.87$) indicating common sources. The good correlation of the esters with CO ($R^2 = 0.74–0.88$) points to an anthropogenic origin related to well-mixed air from the polluted boundary layer. The correlation to CO was weaker in summer (R=0.23–0.51), which most likely was due to photochemical production by the oxidation of alkenes.

3.1.3   Alcohols

In agreement with the previous studies in Table 2, methanol was the alcohol with the highest mixing ratios. The IQR at Jungfraujoch in summer (597–956 ppt) was in between the values from the Arctic (256 ppt) and the values reported over the Pacific (1250 ppt), and in range of the measurements at the west coast of the USA (611–1021 ppt). In the European background air methanol values of about 600 to 2000 ppt and 500 to 1500 ppt were reported from an aircraft campaign at 6 km altitude and at Mace Head, respectively [Colomb et al., 2006; Lewis et al., 2005]. The remote station at Mace Head is, depending on the wind direction, used to study the background marine air or European continental emissions. There are both anthropogenic and biogenic sources of methanol [Schade and Goldstein, 2006]. Biogenic sources of methanol have been summarized in a review by Fall [2003]. Furthermore, methanol is a secondary trace gas, and methane oxidation is postulated to be a substantial source as well [Jacob et al., 2005]. Evidence for secondary production of methanol in biomass burning plumes was reported by Holzinger et al. [2005] from aircraft measurements in southern Europe. Wisthaler et al. [2002] reported mean mixing ratios of methanol from 627 to 1417 ppt over the Indian ocean, depending on the origin of the air mass. Low mixing ratios were observed when the air mass arrived from the southern hemisphere, whereas elevated mixing ratios were seen in air from the continent. In this case also biomass burning was identified as a major source.

Ethanol was present at Jungfraujoch at substantially lower levels than methanol, and had its highest mixing ratios in winter with an IQR of 135–410 ppt. About 50% of ethanol in the background atmosphere over the Pacific was estimated to origin from primary biogenic sources [Singh et al., 2004]. Further, biomass burning, anthropogenic emissions and hydrocarbon oxidation were estimated to explain about 1/6 each. Ethanol at Jungfraujoch was in winter well correlated with CO in polluted air masses. Therefore it is assumed that a substantial part of the ethanol has been emitted by anthropogenic sources.

In winter and spring, the average iso-propanol mixing ratio was 20% of that of ethanol. Sources for this compound are window cleaning fluids [Legreid et al., 2007] and industrial solvents. Propanol and 2-methyl-3- buten-2-ol (MBO) are present at low ppt levels. MBO correlates well with MTBE (R = 0.82) in summer providing evidence for anthropogenic sources, although the sources for MBO described in the literature have only been of biogenic origin [Schade and Goldstein, 2001].

3.1.4 Ketones

Acetone was the most abundant OVOC measured at Jungfraujoch, showing only a moderate seasonal variation with an IQR of 718–991 ppt in summer and 487–749 ppt winter. These values are comparable to other studies in background atmospheres over the Pacific (822 ppt), from the Arctic (871 ppt), at the west coast of USA (529–801 ppt) and over the Indian Ocean (515–2080 ppt) (see Table 2). Similar values were also reported during a summer campaign at Mace Head [Lewis et al., 2005]. In general, mixing ratios of acetone are higher in continental air masses. Apart from its anthropogenic emissions from exhaust and solvent usage, acetone is reported to be emitted by biogenic sources [Fall, 2003; Poschl et al., 2001], biomass burning [Holzinger et al., 1999], and related to secondary production from MBO [Alvarado et al., 1999] or a-pinene [Reissell et al., 1999]. Acetone has been shown to be a globally abundant compound which can act as a source of HOx, peroxy and alkoxy radicals in the free troposphere [Arnold et al., 1997; Singh et al., 1994]. Jacob et al. [2002] estimated the oxidation of anthropogenic iso-alkanes to be the main source of acetone in the northern hemisphere in all seasons except summer. Terrestrial vegetation and oceans were identified as additional sources. The methanol/acetone ratio at Jungfraujoch varies from 0.5 in fall to 1.0 in spring, which is comparable to the study of Karl et al. [2001a] at Sonnblick in Austria (3100 m asl). It is lower than in a 1-year study of Schade and Goldstein [2006] at a rural site in USA with a measured methanol/acetone ratio of about 2 to 5. This could be due to different source patterns between these substances in Europe compared to the USA. In the Arctic the ratio was as low as 0.3 under 24-h solar light conditions [Boudries et al., 2002], whereas the ratio was between 0.6 and 1.5 for other studies at remote sites (Table 2).

Methyl ethyl ketone (MEK) was far less abundant than acetone with a mean mixing ratio of 101 ppt in winter, which is in the range of the values reported from the west coast of the USA (45–76 ppt), the Arctic (55 ppt), and the Pacific (75 ppt) (see Table 2). The seasonal cycle was similar to benzene with higher mixing ratios in winter compared to summer, which implies dominant anthropogenic sources. MEK has several industrial sources [U.S. Environmental Protection Agency (U.S. EPA), 1994], but Riemer et al. [1998] also found evidence for biogenic sources of MEK from the good correlation of MEK with acetone and methanol in the absence of anthropogenic sources at a rural site in southeastern USA. At Jungfraujoch, MEK correlated well with acetone (R = 0.83), methyl acetate (R = 0.85), benzene (R = 0.78) and butane (R = 0.73) in summer. The correlations with the solely anthropogenic emitted benzene and butane suggested significant contributions from anthropogenic sources.

3.1.5 Aldehydes

Formaldehyde was the most abundant aldehyde at Jungfraujoch. Its seasonal cycle with a summer maximum (505 ppt) and lower mixing ratios in spring, fall and winter is explained by elevated photochemical production during the warm season. Grannas et al. [2002] reported an average mixing ratio of 166 ppt in the Arctic boundary layer during sunlight conditions, and measurements performed from a ship in the southern Indian Ocean found 200 ± 70 ppt of formaldehyde [Wagner et al., 2002]. In the European remote marine boundary layer formaldehyde mixing ratios of 130– 430 ppt were found in summer [Still et al., 2006]. The higher mixing ratios at Jungfraujoch can be explained by the influence of the European boundary layer. Formaldehyde mixing ratios up to almost 1000 ppt have been previously measured at the same altitude and latitude as Jungfraujoch during the aircraft campaign TOPSE 2000 [Fried et al., 2003]. Acetaldehyde was the second most abundant aldehyde found in this study, and it showed a summer maximum (IQR: 353–425 ppt) and lowest levels in fall (IQR: 292–331 ppt). These values are somewhat larger than reported from a study over the Pacific (226 ± 89 ppt) [Singh et al., 2004] and from the arctic (166 ± 79 ppt) [Boudries et al., 2002], probably because of the influence from the European boundary layer at Jungfraujoch. At Mace Head acetaldehyde values from about 150 ppt to more than 2000 ppt were reported in summer [Lewis et al., 2005]. The high maximum value at Mace Head can be explained by local anthropogenic and biogenic sources. The observed values at Jungfraujoch were in the range of the mean values reported from the study over the Indian Ocean by Wisthaler et al. [2002] (178–424 ppt), where elevated acetaldehyde mixing ratios within continental air masses implied anthropogenic sources for this compound.

The other aldehydes were present in much lower quantities decreasing with the length of the carbon chain. The propanal mixing ratios in this study were much lower (17–28 ppt), compared to values measured over the Pacific (77 ppt) [Singh et al., 2004], but were somewhat higher than those reported from the Artic (11 ppt) [Boudries et al., 2002]. Butanal was measured at the western coast of USA by Millet et al. [2004] in spring 2002 with an IQR of 15– 23 ppt, which is almost identical to the summer values at Jungfraujoch (15–24 ppt). The unsaturated aldehydes acrolein and methacrolein were also present at Jungfraujoch in the lower ppt range (2–17 ppt and 2–6 ppt, respectively). Acrolein is mainly emitted by combustion of fossil fuel and had therefore higher mixing ratios in winter because of its longer lifetime and more stationary combustion during this season. Methacrolein is a secondary product from the oxidation of isoprene [Biesenthal et al., 1997] and was therefore more abundant in the summer season (2–6 ppt) compared to winter (1–2 ppt).

The best correlation between the aldehydes was found for acetaldehyde and propanal (R = 0.88) in summer, which was also the best correlation found under Artic conditions by Boudries et al. [2002]

6. OVOC measurements at Jungfraujoch

and over the Pacific [Singh et al., 2004]. This could imply similar sources and sinks for the two most abundant compounds among the aldehydes. Furthermore, acetaldehyde correlated also well with acrolein (R = 0.81), acetone (R = 0.80), MEK (R = 0.75) and methanol (R = 0.69).

3.2   Seasonal variation

In Figure 1 the seasonal variation of selected VOCs are compared, showing noteworthy differences between the both primary and secondary OVOCs and the solely primary NMHCs. Several processes contribute to these differences. Benzene and butane have elevated mixing ratios in winter, which can be explained by their prolonged lifetime during this season. This is due to the lower mixing ratio of the OH radical, which is the main factor controlling the atmospheric destruction of these compounds. OVOCs have larger secondary and biogenic sources in summertime, which despite their lower lifetime during this season explains the higher mixing ratios. This trend was also observed at a rural station in California for methanol and acetone [Schade and Goldstein, 2006], but because of their local biogenic sources, their summer median was 5.5 times higher for methanol and 4 times higher for acetone compared to Jungfraujoch. At Jungfraujoch the summer median was 53% and 55% higher than in winter for methanol and acetone, respectively. In addition, the transport from the polluted boundary layer plays an important role, which is studied in more detail in the following.

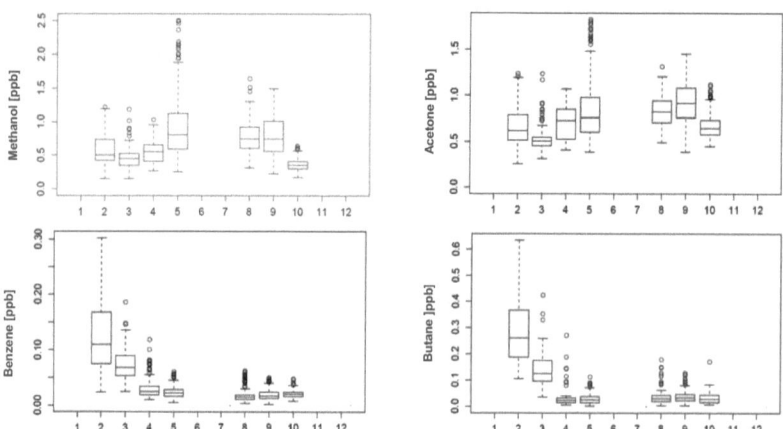

Figure 1. Monthly boxplots during the year 2005 at Jungfraujoch for the compounds methanol, acetone, benzene and butane. The numbers indicate the number of the Calendar months (i.e. 1 for January, 2 for February).

3.3 European source regions

Emissions from the European boundary layer contributed to the mixing ratios measured at Jungfraujoch during periods when polluted air masses are transported to the site by meteorological processes such as front passage and mountain venting [*Henne, et al.*, 2005]. In the following sections the potential source regions of the pollutants are investigated by the methods described in section 2.2 and 2.3.

3.3.1 *From CO/NO_y ratio and visual inspection of back trajectories*

After separating the air masses into undisturbed and disturbed FT conditions (see section 2.2), the air masses qualified as disturbed FT conditions were further separated into northern, southern, eastern and western boundary layer origin by visual inspection of the calculated aLMo trajectories. In Figure 2 a time period in spring is shown, where the origin of the air mass is influenced by the southern boundary layer during 3 days (21.03–24.03). For FT conditions before and afterward the relative humidity and the NO_y/CO ratio was low, as well as the mixing ratios for all organic compounds. It was evident that high NO_y/CO ratios were useful indicators for boundary layer influence. In the morning of 21 May there is a rapid increase in the NO_y/CO ratio, which occurred concurrently with increasing mixing ratios for ozone and the selected OVOCs. The relative humidity was also high during the period. The NO_y/CO ratio and the mixing ratios of ozone and OVOCs decreased in the night and early morning of 23 May as the air masses switched from boundary layer influence to free troposphere air. In the morning of 24 May, the relative humidity reached an unusually low minimum value concurrently with a peaking ozone mixing ratio. This can be explained by stratospheric air masses influencing the concentrations at Jungfraujoch during this event. However, the relatively small enhancement in ozone mixing ratio and the lack of simultaneous changes in the mixing ratios of the OVOCs showed that the stratospheric air was strongly mixed with free tropospheric air before reaching Jungfraujoch.

6. OVOC measurements at Jungfraujoch

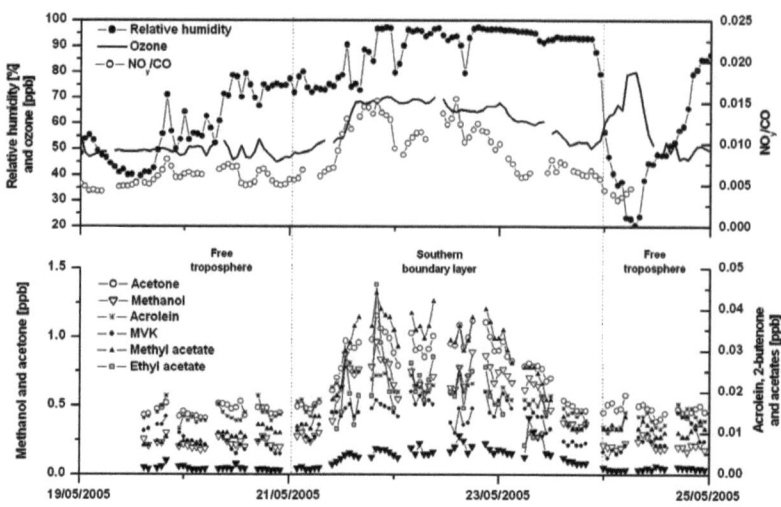

Figure 2. Time series of relative humidity, NO_y/CO ratio and the mixing ratios of ozone (upper panel), and acetone, methanol, acrolein, MVK, methyl acetate and ethyl acetate (lower panel) during 6 days in spring.

Ratios of OVOCs against CO and other anthropogenic tracers have been used either for source allocations [Borbon et al., 2004; de Gouw et al., 2005] or for classification of air masses [De Reus et al., 2003]. For some compounds the ratio to CO was dependent on the origin of the air mass. Figure 3 shows mixing ratios of methyl acetate versus CO mixing ratios at Jungfraujoch for the complete set of data. The data representing northern boundary layer influence at the Jungfraujoch correlated well with CO and showed a 75% enhancement of the slope compared to the free tropospheric conditions. This might be due to emissions from the northern European boundary layer from the exhaust of fossil fuel burning processes. From the east and south there was strong evidence for additional sources not related to CO sources, which can be explained by the use of methyl acetate as an industrial solvent.

6. OVOC measurements at Jungfraujoch

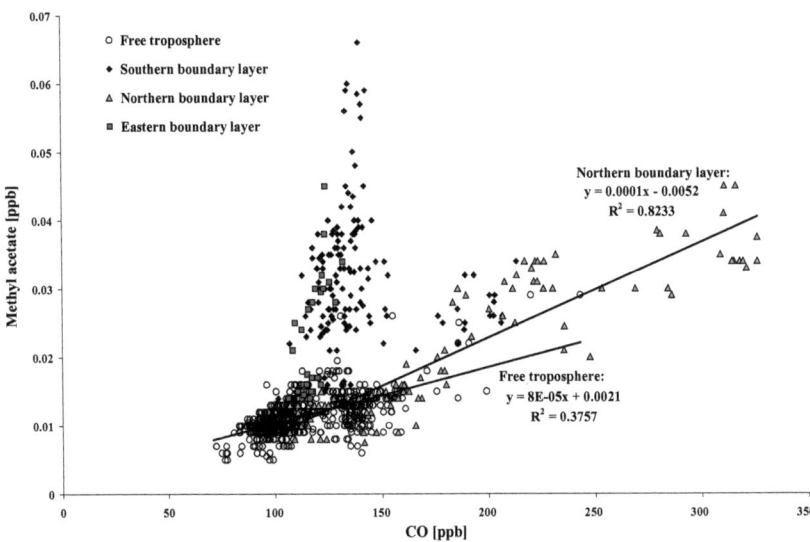

Figure 3. Methyl acetate correlation with CO separated into the source regions "Free troposphere", northern, southern and eastern boundary layer for all data collected.

3.3.2 From statistical trajectory analysis

For a further distinction between European source regions of the OVOCs a more sophisticated approach was applied, which combined the measurement data at Jungfraujoch with statistical trajectory model (section 2.3). The measured concentrations of OVOCs with a lifetime of more than 3 days were combined with concurrent trajectories to retrieve a map of the European source regions. Because of the restricted representativity of the Jungfraujoch site (see section 2.3), the resulting pictures are biased toward central western Europe. They therefore only provide indications of potential source regions. Especially from France the source estimates could be underestimated because of the fact that if air masses are advected to Jungfraujoch from this region, transport is normally fast with low residence time within the boundary layer.

The potential source regions of selected long-lived OVOCs are shown in Figure 4. The trajectory model shows a considerable contribution of methanol and acetone from the south. This seems to be reasonable since these compounds are emitted in large amounts from biogenic sources [Fall, 2003; Jacob et al., 2002, 2005; Singh et al., 1994], which because of the warmer climate are more active in the south of the Alps than in the north. The high contribution from northern Italy and southern France to the measured mixing ratios of methyl acetate and MEK can be explained by the high density of industry in these regions. Methyl acetate is used in glues and nail polish removers, in

6. OVOC measurements at Jungfraujoch

chemical reactions and for extractions (see Figure 4). MEK is a solvent used in the production of paints and coatings [U.S. EPA, 1994]. In the case of ethanol and ethyl acetate there seem to be additional larger sources in Germany and Czech Republic. Ethanol is emitted from combustion sources, fermentation processes and from solvent use in the industry. Ethyl acetate is used in food production as and as a solvent in varnishes and paints.

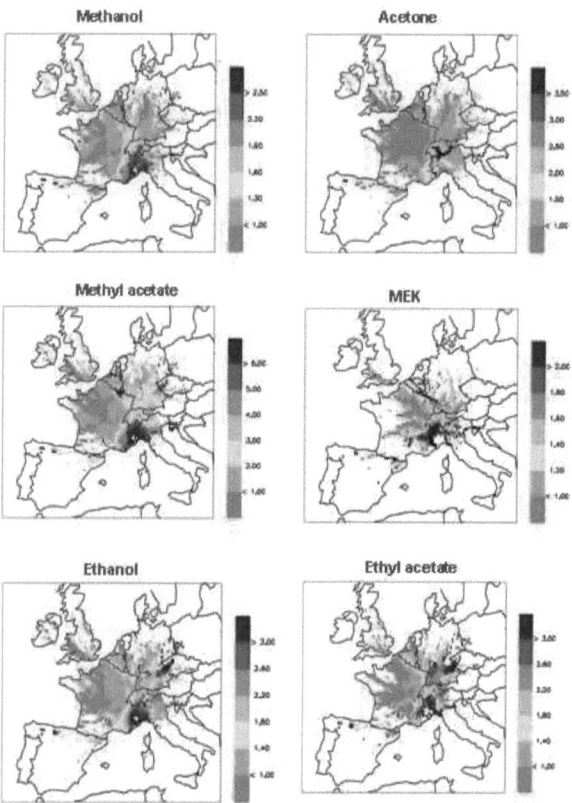

Figure 4. Map of potential source regions of methanol, acetone, methyl acetate, MEK, ethanol and ethyl acetate from about four months of measurement data, as determined by statistical trajectory analysis. Colours give relative importance of regions for measured above baseline mixing ratios at Jungfraujoch. Red colours indicate regions which are, on average, associated with high mixing ratios above baseline measured at Jungfraujoch. Green colours are associated with, on average, small above baseline mixing ratios at Jungfraujoch.

In Figure 5 the results from the trajectory statistics are shown for the two NMHCs butane and benzene. These NMHCs are indicators for anthropogenic sources, and they are at their highest levels when air masses are advected from Germany and from the northern part of Italy. This pattern is also in agreement with source regions identified from halocarbon measurements made at Jungfraujoch [Reimann et al., 2004]. There seems to be strong sources from eastern Germany and the Czech Republic as well, but because of the low number of trajectories from this region the uncertainty is high.

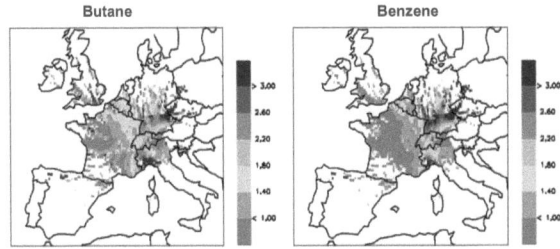

Figure 5. Map of potential source regions of butane and benzene from about four months of measurement data, as determined by statistical trajectory analysis. Colours give relative importance of regions for measured above baseline mixing ratios at Jungfraujoch. Red colours indicate regions which are, on average, associated with high mixing ratios above baseline measured at Jungfraujoch. Green colours are associated with, on average, small above baseline mixing ratios at Jungfraujoch.

4 Conclusions

Four seasonal measurements of OVOCs and selected NMHCs at the high alpine site of Jungfraujoch (Switzerland) have been presented. Most of the OVOCs had highest mixing ratios in summer because of higher emissions from biogenic sources and secondary oxidation processes. On the other hand, anthropogenic NMHCs were more elevated in winter, because of the longer lifetimes and more intensive stationary combustion during this season.

Acetone, methanol, formaldehyde and acetaldehyde were the most abundant compounds measured at the high alpine station in all seasons, and were responsible for 95% of the measured VOCs in summer and 83% in fall. The measured mixing ratios of these compounds were similar to other studies from remote locations. For identification of source regions, a statistical approach using backward trajectories was applied. The anthropogenically emitted compounds butane, benzene, ethanol and ethyl acetate had two main source regions; southern Germany and northern Italy/ southern France, both heavily populated and industrialized areas. For the two industrial solvents methyl acetate

6. OVOC measurements at Jungfraujoch

and butanone (MEK) the main source region was solely northern Italy/southern France. Methanol and acetone, compounds which also have large biogenic sources, had their main contribution from the region south of the Alps as well. This is probably due to the higher biogenic activity in this region compared to the north.

5 Acknowledgements:

This study was financially supported by the Swiss Federal Office for the Environment (FOEN/BAFU). For helpful discussions we thank P.G. Simmonds and D. Young at the University of Bristol. We thank MeteoSwiss for the preparation of meteorological data. We acknowledge the International Foundation High Altitude Research Stations Jungfraujoch and Gornergrat (HFSJG) for making our research possible at the High Altitude Research Station at Jungfraujoch. Last but not least, we thank the custodians at the Jungfraujoch, the Hemunds and the Fischers, for their enthusiasm and patience in supporting our activities.

References:

Alvarado, A., et al. (1999), Products and mechanisms of the gas-phase reactions of OH radicals and O-3 with 2-methyl-3-buten-2-ol, *Atmospheric Environment*, *33*, 2893-2905.

Arnold, F., et al. (1997), Acetone in the upper troposphere and lower stratosphere: Impact on trace gases and aerosols, *Geophysical Research Letters*, *24*, 3017-3020.

Atkinson, R. (2000), Atmospheric chemistry of VOCs and NOx, *Atmospheric Environment*, *34*, 2063-2101.

Balzani Lööv, J. M. (2007), Carbonyls and PANs at Jungfraujoch and the related oxidation processes at the boundary layer/free troposphere interface, thesis, ETH Zurich, Zurich, Switzerland.

Biesenthal, T. A., et al. (1997), A study of relationships between isoprene, its oxidation products, and ozone, in the Lower Fraser Valley, BC, *Atmospheric Environment*, *31*, 2049-2058.

Borbon, A., et al. (2004), Characterising sources and sinks of rural VOC in eastern France, *Chemosphere*, *57*, 931–942

Boudries, H., et al. (2002), Distribution and trends of oxygenated hydrocarbons in the high Arctic derived from measurements in the atmospheric boundary layer and interstitial snow air during the ALERT2000 field campaign, *Atmospheric Environment*, *36*, 2573-2583.

Colomb, A., et al. (2006), Airborne measurements of trace organic species in the upper troposphere over Europe: The impact of deep convection, *Environ. Chem.*, *3*, 244–259.

6. OVOC measurements at Jungfraujoch

de Gouw, J. A., et al. (2005), Budget of organic carbon in a polluted atmosphere: Results from the New England Air Quality Study in 2002, *J. Geophys. Res., 110*, D16305, doi:10.1029/2004JD005623.

De Reus, M., et al. (2003), On the relationship between acetone and carbon monoxide in different air masses, *Atmos. Chem. Phys., 3*, 1709–1723.

Fall, R. (2003), Abundant oxygenates in the atmosphere: A biochemical perspective, *Chemical Reviews, 103*, 4941-4951.

Grannas, A. M., et al. (2002), A study of photochemical and physical processes affecting carbonyl compounds in the Arctic atmospheric boundary layer, *Atmospheric Environment, 36*, 2733-2742.

Henne, S., et al. (2005), Influence of mountain venting in the Alps on the ozone chemistry of the lower free troposphere and the European pollution export, *Journal of Geophysical Research-Atmospheres, 110*.

Holzinger, R., et al. (1999), Biomass burning as a source of formaldehyde, acetaldehyde, methanol, acetone, acetonitrile, and hydrogen cyanide, *Geophysical Research Letters, 26*, 1161-1164.

Holzinger, R., et al. (2005), Oxygenated compounds in aged biomass burning plumes over the Eastern Mediterranean: evidence for strong secondary production of methanol and acetone, *Atmospheric Chemistry And Physics, 5*, 39-46.

Jacob, D. J., et al. (2002), Atmospheric budget of acetone, *Journal Of Geophysical Research-Atmospheres, 107*.

Jacob, D. J., et al. (2005), Global budget of methanol: Constraints from atmospheric observations, *Journal Of Geophysical Research-Atmospheres, 110*.

Junkermann, W., and J. M. Burger (2006), A new portable instrument for continuous measurement of formaldehyde in ambient air, *J. Atmos. Oceanic Technol., 23*, 38–45.

Karl, T., et al. (2001a), Variability-lifetime relationship of VOCs observed at the Sonnblick Observatory 1999—Estimation of HO-densities, *Atmos. Environ., 35*, 5287–5300.

Karl, T., et al. (2001b), High concentrations of reactive biogenic VOCs at a high altitude site in late autumn, *Geophys. Res. Lett., 28*, 507–510.

Komenda, M., et al. (2003), Description and characterization of an on-line system for long- term measurements of isoprene, methyl vinyl ketone, and methacrolein in ambient air, *Journal of Chromatography A, 995*, 185-201.

Legreid, G., J. Balzani Lööv, J. Staehelin, C. Hu"glin, M. Hill, B. Buchmann, A. S. H. Prevot, and S. Reimann (2007), Oxygenated volatile organic compounds (OVOCs) at an urban background site in Zürich (Europe). Seasonal variation and source allocation, *Atmos. Environ., 41*, 8409–8423.

Lewis, A. C., et al. (2005), Sources and sinks of acetone, methanol, and acetaldehyde in North Atlantic air, *Atmos. Chem. Phys., 5*, 1285–1317.

6. OVOC measurements at Jungfraujoch

Millet, D. B., et al. (2004), Volatile organic compound measurements at Trinidad Head, California, during ITCT 2K2: Analysis of sources, atmospheric composition, and aerosol residence times, *Journal Of Geophysical Research-Atmospheres*, *109*.

Poschl, U., et al. (2001), High acetone concentrations throughout the 0– 12 km altitude range over the tropical rainforest in Surinam, *J. Atmos. Chem.*, *38*, 115–132.

Reimann, S., et al. (2004), Halogenated greenhouse gases at the Swiss High Alpine Site of Jungfraujoch (3580 m asl): Continuous measurements and their use for regional European source allocation, *Journal Of Geophysical Research-Atmospheres*, *109*.

Reissell, A., et al. (1999), Formation of acetone from the OH radical- and O-3-initiated reactions of a series of monoterpenes, *Journal Of Geophysical Research-Atmospheres*, *104*, 13869-13879.

Riemer, D., et al. (1998), Observations of nonmethane hydrocarbons and oxygenated volatile organic compounds at a rural site in the southeastern United States, *Journal Of Geophysical Research-Atmospheres*, *103*, 28111-28128.

Schade, G. W., and A. H. Goldstein (2001), Fluxes of oxygenated volatile organic compounds from a ponderosa pine plantation, *Journal Of Geophysical Research-Atmospheres*, *106*, 3111-3123.

Schade, G. W., and A. H. Goldstein (2006), Seasonal measurements of acetone and methanol: Abundances and implications for atmospheric budgets, *Global Biogeochemical Cycles*, *20*.

Seibert, P., et al. (1994), Trajectory analysis of aerosol measurements at high alpine sites, in Transport and Transformation of Pollutants in the Troposphere, *edited by P. M. Borrell et al.*, 689– 693.

Singh, H., et al. (2000), Distribution and fate of selected oxygenated organic species in the troposphere and lower stratosphere over the Atlantic, *J. Geophys. Res.-Atmos.*, *105*, 3795-3805.

Singh, H. B., et al. (1994), Acetone In The Atmosphere - Distribution, Sources, And Sinks, *Journal Of Geophysical Research-Atmospheres*, *99*, 1805-1819.

Singh, H. B., et al. (2004), Analysis of the atmospheric distribution, sources, and sinks of oxygenated volatile organic chemicals based on measurements over the Pacific during TRACE-P, *Journal of Geophysical Research-Atmospheres*, *109*, art. no.-D15S07.

Solberg, S., et al. (1996), Carbonyls and nonmethane hydrocarbons at rural European sites from the Mediterranean to the Arctic, *Journal Of Atmospheric Chemistry*, *25*, 33-66.

Still, T. J., et al. (2006), Ambient formaldehyde measurements made at a remote marine boundary layer site during the NAMBLEX campaign - a comparison of data from chromatographic and modified Hantzsch techniques, *Atmospheric Chemistry and Physics*, *6*, 2711-2726.

Stohl, A. (1996), Trajectory statistics - A new method to establish source-receptor relationships of air pollutants and its application to the transport of particulate sulfate in Europe, *Atmospheric Environment*, *30*, 579-587.

6. OVOC measurements at Jungfraujoch

U.S. Environmental Protection Agency (1994), Locating and estimating air emissions from sources of methyl ethyl ketone, Rep. EPA-454/R-93-046, Off. of Air Qual. Plann. and Stand., Research Triangle Park, N. C.

Wagner, V., R. von Glasow, H. Fischer, and P. J. Crutzen (2002), Are CH2O measurements in the marine boundary layer suitable for testing the current understanding of CH4 photooxidation?: A model study, *J. Geophys. Res., 107(D3)*, 4029, doi:10.1029/2001JD000722.

Walker, S. J., et al. (2006), Processes controlling the concentration of hydroperoxides at Jungfraujoch Observatory, Switzerland, *Atmospheric Chemistry and Physics, 6*, 5525-5536.

Wennberg, P. O., et al. (1998), Hydrogen radicals, nitrogen radicals, and the production of O-3 in the upper troposphere, *Science, 279*, 49-53.

Wisthaler, A., et al. (2002), Organic trace gas measurements by PTR-MS during INDOEX 1999, *Journal of Geophysical Research-Atmospheres, 107*, art. no.-8024.

Zanis, P., et al. (2006), Seasonal variability of measured Ozone production efficiencies in the lower free troposphere of Central Europe *Atmospheric Chemistry and Physics Discussions (ACPD)*, 9315-9349.

Zellweger, C., et al. (2000), Summertime NOy speciation at the Jungfraujoch, 3580 m above sea level, Switzerland, *Journal Of Geophysical Research-Atmospheres, 105*, 6655-6667.

Zellweger, C., et al. (2003), Partitioning of reactive nitrogen (NOy) and dependence on meteorological conditions in the lower free troposphere, *Atmospheric Chemistry And Physics, 3*, 779-796.

7. Conclusions and outlook

An instrument for the measurements of Oxygenated Volatile Organic Compounds (OVOCs) has been successfully developed and applied during twelve measurement campaigns in Switzerland. The analytical system was based on an adsorption desorption system (ADS) originally developed for halocarbon analysis in the Atmospheric Chemistry group at the University of Bristol, UK. The main challenges to be solved were the adsorption of polar compounds in the system, the removal of humidity without sample losses, the artefact formation by ozone, the calibration of the instrument and the reduction of system blank values. The sample analysis was performed on a gas chromatograph-mass spectrometer (GC-MS), which was optimized for the compounds of interest. The accuracy of the system was tested during an intercomparison campaign organized by the European Network ACCENT at the SAPHIR smog-chamber in Jülich, Germany. For most compounds the accuracy was within 1-20 %. The exceptions were the alcohols with lower accuracies of 20-40 %. Due to ozone artefact formation of several aldehydes during the first campaigns, an ozone removal based on the titration with nitrogen oxide was introduced.

The analytical capabilities of the instrument offered the rather unique opportunity to measure a variety of organic trace gases. They include primary compounds emitted from different anthropogenic, as well as biogenic sources, and secondary compounds produced by oxidation processes in the atmosphere. Measured compounds were carbonyls, key intermediates of tropospheric chemistry as well as primary anthropogenic and biogenic compounds, alcohols, emitted by both anthropogenic and biogenic sources and ether and esters, mainly emitted by anthropogenic sources. Furthermore, selected anthropogenic and biogenic NMHCs were measured.

The measurement campaigns were performed at an urban background location in Zürich (Kasernenhof) and at the high alpine station Jungfraujoch in order to gather information about distribution and sources of the OVOCs in Switzerland. These were the first measurements for many of the OVOCs in the Swiss polluted boundary layer, and represent therefore a unique set of data. The OVOC source profiles for the specific sources road traffic and biomass burning for residential heating were in addition retrieved from measurements in a highway tunnel close to Zürich (Gubrist tunnel), and in a village in the Swiss Alps (Roveredo), in which the air was highly influenced by wood burning emissions in winter.

From the measurement campaign in the Gubrist highway tunnel, emission factors (EFs) for OVOCs and selected non-methane hydrocarbons (NMHCs) were retrieved. The compound with the highest EF for the total fleet was ethanol. This compound was not only related to exhaust emissions, but also to the use of window wiper fluid, which resulted in peak mixing ratios of more than 1 ppm. In

7. Conclusions and outlook

total, the OVOCs represented 54 % of the measured volatile organic compounds (VOCs) from mobile sources. The relative contributions of light-duty vehicles (LDV) and heavy-duty vehicles (HDV) to the total emissions indicated that OVOCs were mainly produced by the HDVs, whereas the LDVs dominated the emissions of the NMHCs. The comparison with earlier campaigns at the same site confirmed the continuous long-term decrease of organic exhaust emissions at highway conditions due to steady improvements of vehicle technology.

In Zürich, ethanol was also the dominating compound measured throughout all seasons. Its anthropogenic origin was indicated by higher mixing ratios in winter than in summer, which was also the case for known anthropogenic pollutants like benzene and acrolein. Methanol, acetone, isoprene and other compounds with additional biogenic sources had higher levels during summer compared to winter. By a source-tracer-ratio method with carbon monoxide (CO) as tracer for combustion sources, it was possible to quantify the contribution of the daily accumulated anthropogenic combustion to the mixing ratios of each specific volatile organic compound (VOC) in Zürich. Using this method, it was estimated that the local sources (combustion) explained 40 % (28 %) of the measured OVOC in summer and 49 % (33 %) in winter. The main contribution of both the OVOC and NMHC levels in Zürich were explained by the regional background, which was defined as the daily running 0.1 quantile of the measured values. From the calculation of the incremental ozone production, it was estimated that the total OVOCs explained 40 % of the total VOC ozone production. Local OVOC sources were responsible for 16 %.

The contribution of the large-scale background to the VOC mixing ratios in Zürich was defined as the seasonal 0.1 quantile of measured mixing ratios at Jungfraujoch. Despite the lower lifetimes of the OVOCs in summer, the summertime large-scale background accounted for a higher fraction of the OVOCs in Zürich compared to winter (10% vs. 6%). The higher summertime contribution is could be due to more vertical mixing of air masses and higher biogenic and secondary production of acetaldehyde, acetone and methanol, the most abundant compounds measured the high alpine station Jungfraujoch. At this station they were responsible for 82% of the measured VOCs in summer and 51 % in fall. The measured mixing ratios of these compounds were in accordance with other studies from remote locations (see Table 1). For days with influence from the polluted boundary layer (PBL) a statistical approach using backward trajectories was applied to identify source regions. The anthropogenically emitted compounds butane, benzene, ethanol and ethyl acetate had two main source regions; southern Germany and northern Italy, both heavily populated and industrialized areas. For the two industrial solvents methyl acetate and butanone (MEK) the main source region was solely northern Italy. Methanol and acetone, compounds which also have large biogenic sources, had their

7. Conclusions and outlook

main contribution from the region south of the Alps as well. This is probably due to the higher biogenic activity in this region compared to the north.

The importance of the measured OVOCs to local, regional and global photochemistry has been proven during these campaigns. They are dominant components in the free troposphere as well as in urban regions. For a better source assessment it would be recommendable to perform more measurements in the vicinity of sources. In this study it proved difficult to use the measured profiles from the highway tunnel and the alpine village to retrieve the road traffic and biomass burning (from residential heating) contributions in Zürich. For road traffic this was probably due to other driving conditions with more cold-starts and congestion of traffic in the urban centre resulting in different traffic emission profiles. For the biomass burning this can be explained by many interfering combustion sources with similar patterns in Zürich.

The OVOC measurement data probably contain more valuable information, which should be analyzed further. First trials with positive matrix factorisation (PMF) showed interesting results, and could have the potential to reveal more source information. In general, I propose also to continue source specific measurements to improve the ability to model the source contributions to OVOC mixing ratios in urban, rural and remote regions. This could be measurements in park houses for cold start emissions, in industrial regions and in addition, direct measurements of wood burning emissions in laboratory experiments. The dataset from the OVOC measurements in this PhD thesis should be used for comparison with numerical simulations as well.

7. Conclusions and outlook

Compounds	Zürich Summer Mean	Zürich Winter Mean	Other urban locations – Athens, Greece 2000 Mean (Site1)	Athens, Greece 2000 Mean (Site2)	Rome, Italy 1994/1995 Mean (Jan)	Rome, Italy 1994/1995 Mean (Jun/Jul)	São Paulo, Brazil 1998 Mean	São Paulo, Brazil 1998 St.dev.	Stockholm, Sweden 1982 Mean (Site5)*	Stockholm, Sweden 1982 Mean (Site1)**	Jungfraujoch Summer Mean	Jungfraujoch Winter Mean	High arctic 2000 Mean	High arctic 2000 St.dev.	Trinidad Head, USA 2002 1. Quartile	Trinidad Head, USA 2002 3. Quartile	Pacific 2001 Mean	Pacific 2001 St.dev.
Formaldehyde	2.35	1.83	9.7 - 17.2		11.2 - 17		5.0 ± 2.8		5.79 - 5.29		na	na					0.47 - 0.68	
Acetaldehyde	0.80	0.82	10.6 - 15.1		4.6 - 9.3		5.4 ± 2.8				0.39	na	0.20 ± 0.09				0.37 - 0.42	
Propanal	0.12	0.12	1.0 - 2.0		0.9 - 1.8						0.025	na	0.01 ± 0.01				0.14 - 0.19	
Butanal	0.046	0.041	1.6 - 7.4		1.0* - 1.4*				0.62 - 1.7		0.022	na						
Pentanal	0.030	0.023							0.49 - 1.07		0.011	na						
Hexanal	0.062	0.024									0.011	na						
Benzaldehyde	0.019	0.017	1.7 - 3.2		0.50 - 0.70						0.004	0.002			0.001 - 0.006			
MTBE	0.058	0.142	0.8 - 1.6								0.005	0.008						
Acrolein	0.041	0.017							0.25 - 3.42		0.005	0.001			0.01 - 0.02			
Methacrolein	0.178	0.087							0.06 - 0.71		0.008	0.004			0.53 - 0.80		0.82 - 0.50	
Acetone	2.12	1.17	5.3* - 13.7*						4.3 - 19.4		0.87	0.49	0.48 ± 0.14		0.003 - 0.009			
MVK	0.078	0.034			0.4** - 0.7***						0.013	0.005			0.05 - 0.076		0.13 - 0.15	
MEK	0.20	0.22							3.59 - 8.47		0.037	0.063	0.05 ± 0.04		0.61 - 1.02		1.10 - 1.25	
Methanol	3.18	1.21					34.1 ± 9.4		8.56 - 26.7		0.77	0.40	0.27 ± 0.12		0.08 - 0.17		0.17 - 0.25	
Ethanol	3.94	7.53					176.3 ± 38.1		2.62 - 82.8		0.12	0.13	0.08 ± 0.05		0.01 - 0.03			
iso-propanol	0.51	1.00							0.3 - 14.3		na	0.034						
Propanol	0.027	0.083									0.002	0.002						
Butanol	0.11	0.16									na	na						
MBO	0.066	0.063									0.006	0.004			0.00 - 0.018			
Methyl acetate	0.052	0.075									0.013	0.012						
Ethyl acetate	0.19	0.25							0.23 - 2.64		0.009	0.011						
Butyl acetate	0.060	0.128									0.005	0.003						
			*Acetone + Acrolein		* Butanal+MEK ** MVK+methacrolein				* Urban background ** Road site									

Table 1. Mean values from measurements in the urban background station in Zürich and at the high-alpine station Jungfraujoch. The values are compared with reported values from other studies at similar locations.

Abbreviations

ACCENT	Atmospheric Composition Change the European Network of Excellence
ADS	Adsorption Desorption System
aLMo	Alpine Model
asl.	above sea level
CIR-MS	Chemical ionization reaction – mass spectrometry
FT	Free Troposphere
GC	Gas Chromatograph
GC-MS	Gas Chromatograph – Mass Spectrometer
ID	Internal Diameter
MADS	Modified Adsorption Desorption System
MEK	Methyl Ethyl Ketone (Butanone)
MBO	2-Methyl-3-buten-2-ol
MTBE	Methyl-Tert-Butyl-Ether
MVK	Methyl Vinyl Ketone (2-Butenone)
NABEL	the Swiss air pollution monitoring Network (German)
NMHC(s)	Non-Methane HydroCarbon(s)
OVOC(s)	Oxygenated Volatile Organic Compound(s)
PBL	Polluted Boundary Layer
PFA	PerFluoroAlkoxy
ppb	parts per billion (10^{-9})
ppt	part per trillion (10^{-12})
PTR-MS	Proton Transfer Reaction – Mass Spectrometer
SAPHIR	Simulation of Atmospheric PHotochemistry In a large Reaction chamber
SIM	Selected Ion Monitoring
VOC	Volatile Organic Compound
ZA	Zero Air

Appendix I

As a part of the Aerowood campaign, which was initiated to study the impact of wood combustion on aerosol concentration, OVOCs were measured during two weeks in December 2005. The campaign site was a small town south of the Swiss alps in a narrow valley. There is both a highway passing through the town and about 1000 inhabitants using mostly wood combustion in winter for heating. Table 1 gives an overview of the mixing ratios during the two weeks. Based on the mean, the alcohols were the main compound group measured in Roveredo with 55% of total measured VOCs and OVOCs. The aromatic followed with 20%, then the ketones with 7,7% and the aldehydes with 7.4%. MTBE, solely emitted by gasoline driven vehicles, show a different diurnal cycle than the OVOCs (Figure 1).

Compounds (ppb)	Roveredo - December 2005		
	Median	Mean	IQR
Acetaldehyde	0.315	0.398	0.146 - 0.564
Propanal	0.087	0.108	0.044 - 0.146
Butanal	0.044	0.048	0.026 - 0.061
Pentanal	0.019	0.023	0.013 - 0.031
Hexanal	0.022	0.024	0.016 - 0.031
Benzaldehyde	0.012	0.019	0.005 - 0.028
Acrolein	0.283	0.395	0.134 - 0.537
Methacrolein	0.033	0.039	0.015 - 0.053
MTBE	0.096	0.122	0.061 - 0.160
Acetone	0.756	0.824	0.485 - 1.047
MVK	0.061	0.086	0.029 - 0.118
MEK	0.169	0.191	0.094 - 0.245
Methanol	1.179	1.507	0.594 - 1.971
Ethanol	2.986	4.759	1.779 - 6.243
Iso-propanol	0.691	1.159	0.379 - 1.352
n-Propanol	0.015	0.020	0.007 - 0.030
2-Methyl-3-buten-2-ol	0.069	0.096	0.046 - 0.120
n-Butanol	0.104	0.342	0.036 - 0.430
Methyl acetate	0.280	0.350	0.164 - 0.476
Ethyl acetate	0.045	0.055	0.027 - 0.074
Butyl acetate	0.029	0.050	0.016 - 0.054
Butane	0.475	0.546	0.358 - 0.660
1,3-Butadiene	0.152	0.175	0.080 - 0.239
Isoprene	0.049	0.055	0.025 - 0.076
Benzene	0.632	0.724	0.385 - 0.953
Toluene	0.972	1.172	0.590 - 1.545
Ethylbenzene	0.110	0.138	0.062 - 0.183
m,p-Xylene	0.391	0.504	0.211 - 0.659
o-Xylene	0.139	0.179	0.077 - 0.231
1,3,5-Trimethylbenzene	0.064	0.084	0.037 - 0.104
1,2,4-Trimethylbenzene	0.085	0.110	0.050 - 0.133

Table 1. Median, mean and interquartile range (IQR) for the measured OVOCs and VOCs in Roveredo in December 2005.

Appendix I

In Figure 1 the diurnal cycles for selected compounds are shown. Benzene has a similar cycle as MTBE, indicating that traffic is the main source for this compound.

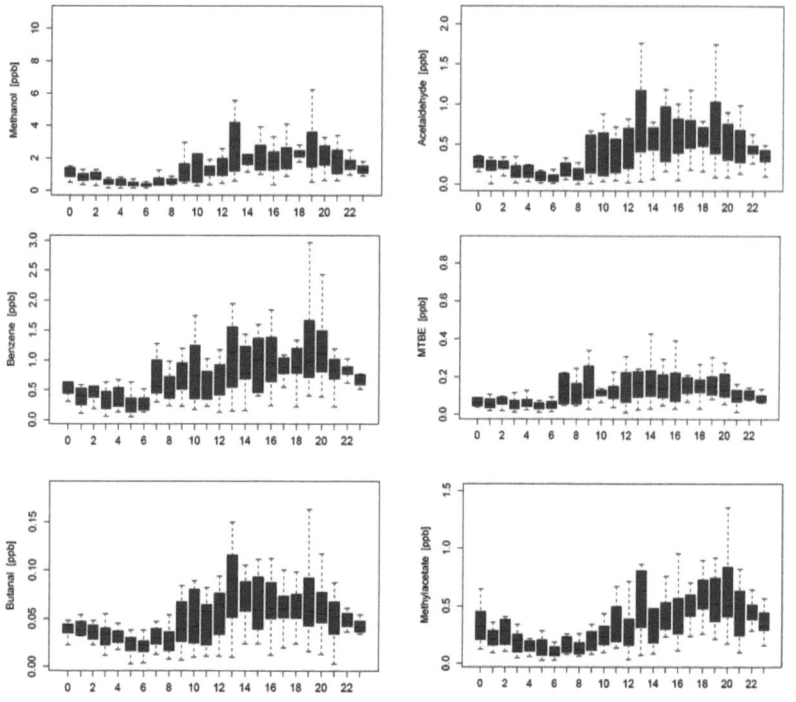

Figure.1. Diurnal cycles of methanol, acetaldehyde, benzene, MTBE, butanal and methyl acetate during two weeks of measurement in Roveredo – December 2005.

Appendix II

Table 1 shows the gas standard mixtures which were applied during the various experiments and measurement campaigns.

Compounds [ppb]	Empa standard	Bristol standard	FAL standard	NPL standard	New empa standard	PSI standard	New NPL standard
Methylether	411				401		
Methanol		522	807		394		
Butane	408	548		1.67	401		4.01
1,3-Butadien				4.8			4.05
Acetaldehyde	407	453	1000		404	1010	
Ethanol	406	542	1000		394		
Isoprene			200	2.58			4.01
Acrolein	394	468			385	195	
Propanal	384	480			414		
Methyl acetate	403				391		
Iso-propanol	408				400		
Acetone	415	536	200		406	215	
n-Propanol	407	532			399		
MTBE	404				407		
Methacrolien	390	412			406		
Benzene	402	516		4.74	403		4.04
Ethyl acetate	407				401		
Butanal	337	504			400		
MVK	390	511			371	190	
Iso-butanol	390				393		
MEK	414	510			403		
2-Methyl-3-buten-2-ol	413	540			404		
n-Butanol	394	491			395		
Toluene			200	3.67	403	210	3.99
n-Pentanal	340	481			398		
Butyl acetate	411				401		
Ethylbenzene				1.5			4.13
m+p-Xylene				1.86			8.16
Hexanal	373	476			373		
o-Xylene				1.23			4.05
Styrene							
1-ethyl-2-methyl-							
1-ethyl-3-methyl-				2.04			
1,2,4-trimethylbenzene				0.94		200	4.15
Benzaldehyde	405				404	110	
Decane							
Nonanal					401		
a-pinene			200				
Propene						202	4.05

Table 1. Overview over concentrations in the used gas standard mixture in the experiments and measurement campaigns during this PhD project.

Acknowledgements

Through fourteen (!) measurement campaigns I've got to know a few people during my PhD work. I feel privileged to have been working with so many devoted scientists. I thereby learned a lot in these three years, and have a lot of people to express my gratitude to: First of all, my thanks go to Dr. Stefan Reimann, for he was the one who convinced to start this PhD, guided me through the work, was always motivating through his positive attitude and taught me in his logical style the key issues of air measurements and data interpretation. Also, special thanks to Prof. Johannes Staehelin for being my supervisor and for his great support through the last months of the PhD. I wish to express my gratitude to the Atmospheric Chemistry group in Bristol, UK, for teaching me more about construction of air sample devices and for the great time working with them in Bristol. Dickon Young deserves special mention for all fruitful discussions and problem solving, for all good time spent together at various measurement sites, and last but not least for providing me with some of his data for this PhD thesis

I'm grateful for the collegial support of Konrad Stemmler, Martin Steinbacher, Martin Vollmer and the other colleagues at Empa. Also to the people in the Laboratory for Atmospheric Chemistry at Paul Scherrer Institute, especially Andre Prevot, Josef Dommen and Astrid Gascho for the interesting time I spent at their Institute. A special thanks goes to Jacob Balzani Lööv from ETH Zürich for his support and for the time spent together during the campaigns at Jungfraujoch.

Thanks to everyone who was involved in the measurement campaigns, and to the custodians Hemunds and Fischers at Jungfraujoch for the help and the unforgettable time spent at the Jungfraujoch research station

Finally, I want to thank my wife, Mojgan, for her love and encouragement through difficult times, and for her unlimited belief in me.

i want morebooks!

Buy your books fast and straightforward online - at one of world's fastest growing online book stores! Environmentally sound due to Print-on-Demand technologies.

Buy your books online at

www.get-morebooks.com

Kaufen Sie Ihre Bücher schnell und unkompliziert online – auf einer der am schnellsten wachsenden Buchhandelsplattformen weltweit! Dank Print-On-Demand umwelt- und ressourcenschonend produziert.

Bücher schneller online kaufen

www.morebooks.de

VDM Verlagsservicegesellschaft mbH
Heinrich-Böcking-Str. 6-8
D - 66121 Saarbrücken

Telefon: +49 681 3720 174
Telefax: +49 681 3720 1749

info@vdm-vsg.de
www.vdm-vsg.de

Printed by Books on Demand GmbH, Norderstedt / Germany